ROUTLEDGE LIBRARY EDITIONS:
GEOLOGY

Volume 6

ECONOMIC AND APPLIED GEOLOGY

ECONOMIC AND APPLIED GEOLOGY

An Introduction

W.G. SHACKLETON

Routledge
Taylor & Francis Group
LONDON AND NEW YORK

First published in 1986 by Croom Helm Ltd

This edition first published in 2020
by Routledge
2 Park Square, Milton Park, Abingdon, Oxon OX14 4RN

and by Routledge
605 Third Avenue, New York, NY 10017

Routledge is an imprint of the Taylor & Francis Group, an informa business

British Library Cataloguing in Publication Data
A catalogue record for this book is available from the British Library

ISBN 13: 978-0-367-18559-6 (Set)
ISBN 13: 978-0-367-20719-9 (Volume 6) (hbk)

Publisher's Note
The publisher has gone to great lengths to ensure the quality of this reprint but points out that some imperfections in the original copies may be apparent.

Disclaimer
The publisher has made every effort to trace copyright holders and would welcome correspondence from those they have been unable to trace.

Economic and Applied Geology

An Introduction

W.G. SHACKLETON

CROOM HELM
London • Sydney • Wolfeboro, New Hampshire

Croom Helm Ltd, Provident House, Burrell Row,
Beckenham, Kent, BR3 1AT
Croom Helm Australia Pty Ltd, Suite 4, 6th Floor,
64-76 Kippax Street, Surry Hills, NSW 2010, Australia

British Library Cataloguing in Publication Data
Shackleton, W.G.
An introduction to economic and applied
geology.
1. Geology, Economic
I. Title
553 TN260

ISBN 0-7099-3387-8
ISBN 0-7099-4440-3 Pbk

Croom Helm, 27 South Main Street,
Wolfeboro, New Hampshire 03894-2069, USA

Library of Congress Cataloging in Publication
Data applied for:

Printed and bound in Great Britain by
Biddles Ltd, Guildford and King's Lynn

CONTENTS

List of Tables
List of Figures
Introduction

1. **MINERAL DEPOSITS** 1

Classification of Mineral Occurrences.... 1
Resources, Reserves and Ore............ 1
Traditional Classification Schemes..... 3
The Formation of Mineral Deposits........ 4
Igneous Mineral Deposits............... 5
Sedimentary Mineral Deposits........... 10
Residual Deposits...................... 11
Secondary Enrichment................... 12
Weathering............................. 14
Conceptual Models of Mineral Deposit Formation............................ 14

2. **COAL** 19

Occurrence............................... 19
Formation................................ 21
Accumulation........................... 21
Coalification.......................... 22
Constituents............................. 23
Macerals............................... 23
Lithotypes............................. 23
Classification........................... 25

3. **PETROLEUM** 28

Formation................................ 29
Organic Origin......................... 29
Kerogen................................ 31
Petroleum Generation................... 32

	Migration	33
	Occurrences of Petroleum	35
	Petroleum and Geologic Time	35
	Surface Occurrences	35
	Subsurface Occurrences	36
	Extraction	37
	Petroleum Engineering	38
4.	**GROUNDWATER**	40
	Origin of Groundwater	41
	Meteoric Water	41
	Connate Water	41
	Juvenile Water	41
	Storage	42
	Porosity	42
	Effective Porosity	42
	Specific Yield	43
	The Water Table	43
	Groundwater Pressure	44
	Movement of Groundwater	46
	Permeability	46
	Perviosity	47
	Transmissibility	47
	Locating Groundwater	47
	Well Development	50
5.	**RESOURCE EXPLORATION**	52
	Planning	54
	Scanning	56
	Initial Follow-Up	59
	Detailed Follow-Up	60
	Other Targets	60
6.	**EXPLORATION GEOPHYSICS**	62
	Gravity	63
	Magnetic	69
	Seismic	74
	The Seismic Refraction Method	76
	The Seismic Reflection Method	78
	Electrical	80
	Equipotential Line Method	82
	Resistivity Method	83
	Induced Polarisation Method	86
	Electromagnetic Method	89
	Self Potential Method	91
	Radiometric	92

7. **EXPLORATION GEOCHEMISTRY** 95

Mobility.................................. 96
Dispersion................................ 97
Geochemical Statistical Terminology...... 98
Types of Surveys.......................... 101
  Orientation Surveys..................... 101
  Overburden Surveys...................... 101
  Stream Sediment Surveys................. 103
  Vegetation Surveys...................... 104
  Rock Surveys............................ 105
  Water Surveys........................... 105
  Vapour Surveys.......................... 105
  Isotope Surveys......................... 106
Analytical Methods........................ 107

8. **DRILLING TECHNIQUES** 112

Principles................................ 112
Cable or Churn Drills..................... 114
Hammer Drill.............................. 115
Auger Drilling............................ 117
Rotary Drilling........................... 117
Diamond Drilling.......................... 119

9. **BORE HOLE LOGGING** 124

Geological Logging........................ 124
  Cuttings................................ 125
  Core.................................... 126
Geochemical Logging....................... 127
Geophysical Logging....................... 128
  Electric Logs........................... 129
  Radiometric Logs........................ 130
  Sonic Logs.............................. 130
  Magnetic Logs........................... 132
Quantitative Logs......................... 133
Mechanical Logs........................... 133

10. **EXTRACTION TECHNIQUES** 135

Selection of Extraction Technique........ 137
Reserve Calculations...................... 138
Surface Mining............................ 140
  Placer Mining........................... 140
  Open-Cut Mining......................... 142
Underground Mining........................ 146
  Horizontal Bedded Deposits.............. 146
  Lenticular or Massive Deposits......... 147
Fluid Extraction Techniques.............. 150
Solid Extraction by Fluid Techniques..... 153

Environmental Impact of Mining.......... 155

11. MINERAL PROCESSING 159

Preparation............................ 160
Screening............................ 160
Classifying.......................... 160
Comminution.......................... 160
Separating, Sorting or Concentration..... 164
Radiation Sorting.................... 164
Magnetic Sorting..................... 165
Electrical Sorting................... 166
Gravity Sorting...................... 167
The Froth Flotation Process.......... 170
Dewatering........................... 174
Carbon-In-Pulp......................... 175
Design and Assessment of the Process..... 175

12. ENGINEERING GEOLOGY 179

Engineering Properties of Rocks.......... 179
Compressive Strength................. 179
Tensile Strength..................... 180
Engineering Properties of Soils.......... 180
Atterberg Limits..................... 180
Compressibility...................... 181
Shear Strength....................... 182
Soil Water............................. 182
Geotechnical Investigations.............. 183
Foundations for Buildings................ 184
Footings............................. 185
Pile Foundations..................... 186
Settlement............................. 186
Slope Stability.......................... 187
General Excavation Problems.............. 190
Dams..................................... 191
Pavement Foundations..................... 196

13. ENVIRONMENTAL GEOLOGY 199

Volcanic Hazards......................... 199
Seismic Hazards.......................... 200
Health................................... 202
The Natural Environment.............. 202
Environmental Pollution.............. 203
Agriculture.............................. 204
Disposal of Wastes....................... 204

14. GEOLOGY, ECONOMICS AND POLITICS 208

Quantity of Earth Resources.............. 208
Growth Rate of Demand for Earth Resources 209

Index..................................... 215

TABLES AND FIGURES

Tables

7.1 Approximate Median Abundances of Some Elements in Rocks (ppm)............... 111
10.1 A Simple Classification of Extraction Techniques............................ 136
11.1 Distribution of Minerals at Various Positions in Figure 11.26............. 176

Figures

1.1 Illustrations of Epigenetic, Syngenetic, Stratabound and Stratiform Mineralisation........................ 3
1.2 Simplified Diagram of the Bushveld Igneous Complex...................... 5
1.3 Mineral Zoning Associated with the Cornwall Granites..................... 7
1.4 Ideal Porphyry Copper Deposit........... 9
1.5 Possible Locations of High Density Mineral Deposits in a Stream.......... 10
1.6 Generalised Section Through the Weipa Bauxite Deposit...................... 12
1.7 Secondary Enrichment of a Mineral Deposit.............................. 13
1.8 Mineral Deposits and Plate Tectonics.... 15
2.1 A Series of Cyclothems.................. 20
2.2 The Coal Series......................... 24
2.3 A Typical Banded Coal................... 26
2.4 Cannel Coal............................. 26
2.5 The Relationship of Humic Coals to Shales and Sapropelites............... 26
3.1 The Molecular Structures of the Three Main Types of Petroleum............... 29
3.2 Composition, Classification and

Tables and Figures

Evolutionary Paths of Crude Oils...... 30
3.3 Evolutionary Paths of Kerogen Types and Coal Maceral Groups.................. 33
3.4 Primary and Secondary Migration of Petroleum............................ 34
3.5 Quantity of Petroleum Versus the Age of the Host Sediments................... 35
4.1 A Simplified Hydrologic Cycle........... 40
4.2 Some Types of Porosity.................. 42
4.3 Water Table Positions and Zones of Saturation.......................... 43
4.4 Perched Water Tables.................... 44
4.5 Groundwater Under Artesian and Sub-Artesian Conditions.................. 45
4.6 The Effect of Development on the Drawdown Curve of a Well.............. 49
4.7 Drawdown Curve Resulting in Well Pollution............................ 49
4.8 The Johnson Well Screen................. 50
5.1 Time, Cost, Risk and Exploration, Development, Mining.................. 53
5.2 South Australian Mineral Exploration Licences............................ 57
6.1 Bouguer Anomalies and Density Contrasts 65
6.2 Bouguer Anomaly and Residual Anomaly Profiles............................ 66
6.3 The Shape of a Gravity Anomaly and Depth to a Point Source.............. 66
6.4 Different Geological Environments Producing an Identical Gravity Profile 68
6.5 The Magnetic Field Around a Bar Magnet.. 70
6.6 The Earth's Magnetic Field.............. 71
6.7 The Shape of a Magnetic Anomaly Over a Point Source......................... 73
6.8 The Anomaly Resulting from a Vertical Magnetic Body........................ 73
6.9 The Anomaly Resulting from an Inclined Magnetic Body........................ 74
6.10 Reflected and Refracted Waves in a Two Layered Environment.................. 75
6.11 The Path of the Wave at the Critical Angle............................... 76
6.12 The Principles of a Seismic Refraction Survey.............................. 76
6.13 Time-Distance Graph for a Seismic Refraction Survey in a Two Layer Case 77
6.14 The Principles of the Seismic Reflection Survey.............................. 78
6.15 An Example of Multiple Reflections...... 79
6.16 Lines of Current Flow and Equipotential in an Homogenous Conductor............ 82

6.17 The Effects of a Body of High Conductivity on Lines of Current Flow and Equipotential........................ 83
6.18 The Wenner Electrode Configuration...... 84
6.19 The Wenner Expanding Electrode System... 84
6.20 The Results of a Wenner Expanding Electrode Survey...................... 85
6.21 The Schlumberger Electrode Configuration 85
6.22 The Principles of the Induced Polarisation Method......................... 86
6.23 Decay Curves for Different Types of Polarisation.......................... 87
6.24 Dipole-Dipole Induced Polarisation Survey.............................. 88
6.25 Principles of the Electromagnetic Method.............................. 89
6.26 Phase Relationships in an EM System..... 90
6.27 Results of an EM Survey................. 91
6.28 The Principles of the Spontaneous Potential Method..................... 92
7.1 The Geochemical Cycle................... 95
7.2 The Relative Degrees of Dispersion of Zinc, Copper and Lead................ 96
7.3 Primary and Secondary Dispersion........ 97
7.4 The Frequency Distribution of Two Populations of Geochemical Data....... 98
7.5 Regional and Local Background Values.... 100
7.6 Size of Anomaly and Sampling Density.... 101
7.7 Geochemical Anomalies in Transported Soils................................ 102
7.8 The Development of a Dispersion Train... 103
7.9 Representation of Stream Sediment Survey Results by a Worm Diagram............ 107
7.10 Accuracy and Precision.................. 109
8.1 A Typical Rotary Drilling Rig........... 113
8.2 A Cable Tool Drilling Rig............... 115
8.3 A Self-Propelled Crawler Percussion Drill................................ 116
8.4 A Tractor-Mounted Auger Drill........... 117
8.5 Diamond Drill Coring.................... 120
8.6 Field Testing a Diamond Drill Core-Barrel.......................... 121
9.1 The Relationship Between Gamma Ray and Sonic Logs and Lithology............. 131
9.2 The Relationship Between Density, Acid Soluble Iron and "Degree of Magnetism"........................... 132
10.1 Mining Methods for Near Surface Deposits and Deep Deposits............ 138
10.2 Assay Values Across Mineralisation...... 139
10.3 Hypothetical Block of Ore............... 140

10.4 Mechanical Dredge........................ 141
10.5 Loading Gravel from a Pit............... 142
10.6 Strip Mining by Back Casting............ 143
10.7 Multi-Bench Lateral Advance Mining...... 143
10.8 Bowman's Coal Trial Conical Pit, South Australia............................ 144
10.9 Mining the Riocin Stratabound Base Metal Deposit, Spain.................. 145
10.10 Open Pit Mining of Kuroko Ore, Japan.... 146
10.11 Room and Pillar Mining.................. 147
10.12 Longwall Mining......................... 147
10.13 Underground Mining Terminology.......... 148
10.14 Sub-Level Stoping....................... 149
10.15 Cut and Fill Mining..................... 150
10.16 Off-Shore Petroleum Production Platform. 151
10.17 Increased Permeability by Fracturing.... 152
10.18 The Frasch Process...................... 153
10.19 Copper Recovery from Mine Waters, NSW... 154
10.20 "Moonscape" in Cornwall................. 155
10.21 New Use for Old Mine.................... 156
11.1 The Operation of a Jaw Crusher.......... 161
11.2 Locked Middlings........................ 161
11.3 The Operation of a Ball Mill............ 162
11.4 The Interior of the Mill, Arintiero Copper Mine, Spain.................... 163
11.5 Size Frequency Diagrams................. 163
11.6 Radiation Sorting....................... 164
11.7 Magnetic Removal of Tramp Iron.......... 165
11.8 Low Magnetic Intensity Separator........ 166
11.9 Electrostatic Separation................ 166
11.10 Electrodynamic Separation............... 167
11.11 Separation by Density and Size.......... 168
11.12 The Principles of Flowing Film Concentration......................... 169
11.13 Slotted Sluice Box Separation of Heavy Mineral Beach Sand.................... 169
11.14 The Humphries Spiral.................... 170
11.15 Shaking Table Concentration............. 171
11.16 A Froth Flotation Cell.................. 171
11.17 The Structure of a Collector............ 172
11.18 The Relationship Between Contact Angle and Wettability...................... 172
11.19 Concentration of Collector and Contact Angle............................... 172
11.20 Chain Length of Collector and Contact Angle............................... 173
11.21 Relationship Between Grade and Recovery in a Froth Flotation Cell............. 174
11.22 Simple Froth Flotation Circuit.......... 175
11.23 Simple Processing Plant Flow Sheet...... 176
12.1 Determination of Compressive Strength... 180

12.2 Determination of Tensile Strength....... 180
12.3 Atterberg Consistency Limits........... 181
12.4 The Relationship Between Time and Compression........................... 182
12.5 The Flow Net Through an Earth Dam....... 183
12.6 Site Investigation for a Dam........... 184
12.7 Individual and Continuous Footings...... 185
12.8 Bevelled Pier Footings in Expansive Clays................................ 185
12.9 Pile Foundations........................ 186
12.10 The Failure of a Slope................. 187
12.11 Drainage of a Potential Slip Surface.... 188
12.12 A Typical Retaining Wall................ 188
12.13 Artificial Toe, Folkestone Warren England................................. 189
12.14 Reinforced Earth Wall................... 190
12.15 Excavation Side-Wall Failure............ 190
12.16 Well Point Drainage of an Excavation.... 191
12.17a The Myponga Concrete Arch Dam, South Australia............................. 192
12.17b Forming the Asphalt Core, East Dam, High Island Reservoir, Hong Kong..... 193
12.18 A Grout Curtain......................... 194
12.19 Unstable Abutment for an Arch Dam....... 194
12.20 Placing Rip-Rap, West Dam, High Island Reservoir, Hong Kong................. 196
12.21 The Structure of Pavements.............. 196
13.1 Maori Village Partially Buried by Ash... 200
13.2 Public Building Badly Damaged by the 1978 Mekering Earthquake, WA.......... 201
13.3 A Generalised Dose Response Curve....... 202
13.4 Low-Level Radioactive Waste Dump, Radium Hill, South Australia.......... 205
14.1 Ore Grade and Energy Requirements in Copper Production..................... 209
14.2 Copper, Lead and Zinc Prices Deflated by US Wholesale Price Index........... 211
14.3 Weekly Variations in LME Copper Prices, 1983.......................... 212

To Mum and Dad

INTRODUCTION

Economic geology deals with geological materials that are of economic value to our society. These materials include metals and non-metals, coal, petroleum and water. This definition implies the incentive of financial return. However, economic geology is only one part of applied geology which also includes solving engineering and water supply problems; explanation of health and agricultural problems resulting from geological factors; consideration of geological hazards such as earthquakes and volcanism; and suggesting methods of disposal of industrial wastes.

A principal aim of most geology courses is to develop an understanding and knowledge of all aspects of geology. Generally, "pure" or "academic" geology is emphasised. Geology courses in Schools of Mines and similar institutions naturally contain a higher proportion of applied geology although the "basics" of geology still form the bulk of the courses.

Most geology units offered by Colleges of Advanced Education and Polytechnics are included in the courses of training of teachers, national park rangers and similar professionals. The proportion of time allocated for academic subjects in such vocational courses is generally less than in universities. However, a satisfactory understanding of how geological knowledge is used in and by society is very necessary in such vocations, particularly teaching.

According to Knill (1978) the two main industries employing geologists are those concerned with extraction and construction. The extractive industries include those devoted to metallic and non-metallic minerals, construction materials, energy resources, such as coal and petroleum, and water. The construction industries involve the application of geology to civil engineering, development of water resources, transport and foundation engineering. An increasing number of geologists are also employed in the fields of health, agriculture and environmental studies including the disposal of wastes.

It is hoped that the concerned layperson, the senior year high school student and the tertiary undergraduate will find this work a useful introduction to the many ways in which a knowledge of geology can be valuable. As a result, an increasing number of people will understand the geological principles that are the basis of many decisions made today. This will allow them to make decisions based on knowledge rather than hearsay and rumour and be aware of the consequences of those decisions.

Although there are many texts which deal with individual fields of economic and applied geology, until recently there have been few general introductory texts on this subject. A notable exception is Watson (1983). Many introductory and general geology texts also treat economic and applied geology briefly. Clark and Cook (1983), however, devote twenty per cent of their book to this important range of topics.

This book is based on a series of lectures at the Salisbury Campus of the South Australian College of Advanced Education. The book is divided into three major sections. The first deals with the formation and description of different types of geological resources; the second with the exploration, extraction and treatment of these resources; and the third with environmental geology and the impact of applied geology on our civilisation.

The material contained in this book was accumulated from the early 1960s from diverse sources including lecture material, text books, journal articles, symposia and my personal experiences. Due acknowledgement is made to published material where the source is known; if there are any omissions I beg forgiveness.

I particularly thank Barry Cook and Ian Clark for the time and effort they spent reviewing and editing each chapter as it was completed (although any errors and omissions are mine!). Lastly, I thank Mei, Gillian and Fiona for their patience and understanding during the preparation of this book.

REFERENCES

Clark, I. F. and Cook, B. J. (Eds.), 1983. *Geological Science: Perspectives of the Earth*. Australian Academy of Science, Canberra.

Knill, J. L. (Ed.), 1978. *Industrial Geology*. Oxford University Press, Oxford.

Watson, J., 1983. *Geology and Man*. Allen and Unwin, London.

Chapter One

MINERAL DEPOSITS

INTRODUCTION

This chapter, together with those on coal, petroleum and groundwater, provides the background for the rest of this book. "Mineral deposits" is used here in the restrictive sense of applying to naturally formed inorganic metallic and non-metallic deposits. However, the term is often applied to any economic deposit including coal and petroleum. Texts such as Bateman (1950), Stanton (1972) and Park and MacDiarmid (1975) should be used to "flesh out" the following necessarily brief treatment of mineral deposits. References are also made to a number of papers in journals.

CLASSIFICATION OF MINERAL OCCURRENCES

Resources, Reserves and Ore

Mineral occurrences range in size from the very large aggregate we call the Earth to relatively small economically exploited mineral deposits.

The term "Resource" has been defined in many ways, generally implying something in reserve. The first part of this book is concerned with non-renewable resources such as copper, coal and petroleum. Skinner (1976, p13) defines a mineral resource as "presently or potentially extractable concentration of naturally occurring solid, liquid or gaseous material". Note that this is a very broad definition. McDivett and Manners (1974, p73) define a resource as "material that can be expected to become part of the reserve category during the foreseeable future through discovery or through changes in the economic, technological or political conditions".

Reserves, with their economic implications,

are the subject of considerable debate and warrant further discussion. The term "mineral deposit" must first be defined. A mineral deposit can be considered to be a volume of the Earth's crust which contains a mineral or group of minerals in above average concentrations - that is, a positive geochemical anomaly. Traditionally, a mineral deposit from which the required material can be extracted at a profit is termed an ore deposit. However, more recent definitions are somewhat more liberal in their scope. For example, the Australasian Institute of Mining and Metallurgy (Aus. I. M. M.) has defined ore as "a solid naturally occurring aggregate from which one or more valuable constituents may be recovered and which is of sufficient economic interest to require estimation of tonnage and grade" (Aus. I. M. M., 1981).

The same body has defined various categories of ore, given in full below:-

Proved Ore Reserves are those in which the ore has been blocked out in three dimensions by excavation or drilling, but include in addition minor extensions beyond actual openings and drill holes, where the geological factors that limit the ore body are definitely known and where the chance of failure of the ore to reach these limits is so remote not to be a factor in the practical planning of the mine operations.

Probable Ore Reserves cover extensions near at hand to proved ore where the conditions are such that ore will probably be found but where the extent and limiting conditions cannot be so precisely defined as for proved ore. Probable ore reserves may also include ore that has been cut by drill holes too widely spaced to ensure continuity.

Possible Ore (NOT RESERVES) is that for which the relation of the land to adjacent ore bodies and the geological structures warrant some presumption that ore will be found, but where the lack of exploration and development data precludes its being classed as probable.

Synonyms for the above categories are "measured", "indicated" and "inferred", respectively. A more recent discussion on ore reserves is given by King and Others (1982). The techniques involved in the quantitative estimation of mineral resource deposits are discussed in a later section.

The above classifications were prepared for metallic and solid non-metallic deposits but, with some modification, are also used for coal deposits. The actions of a very few unscrupulous geologists and directors of mineral exploration companies in the past several decades has led to the increased control over the issue of reports of a geological nature by stock exchange listed mining or mineral exploration companies. In Australia, for example, such reports must be compiled by a corporate member of the Aus. I. M. M. A readable account of some of the events which led to this requirement in Australia is the book "The Money Miners" (Sykes, 1978).

Traditional Classification Schemes

The early classification schemes of the middle nineteenth century were based on form and genesis. Later that century and to the present time most classification schemes are based mainly on genesis with some minor element of form. The initial main subdivisions of mineral deposits were primary, or bedrock, deposits and secondary, or disintegration, deposits. The terms epigenetic and syngenetic were introduced early this century. Epigenetic deposits are those which formed at a later time than that of the host rock. Syngenetic deposits formed at the same time as the host rock (Figure 1.1).

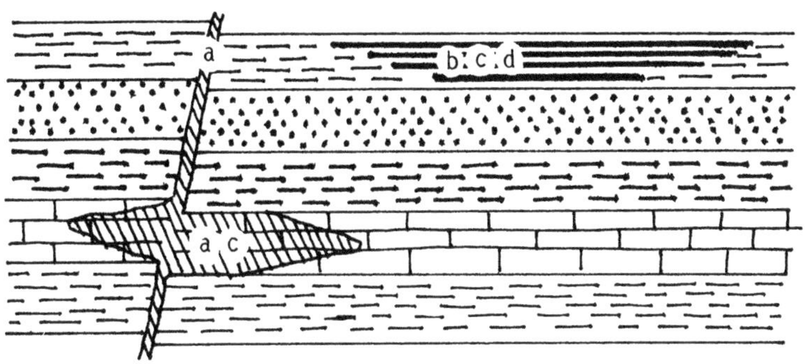

Figure 1.1: Illustrations of (a) Epigenetic, (b) Syngenetic, (c) Stratabound and (d) Stratiform Mineralisation.

Lindgren's classification of 1911 was the first comprehensive genetic classification scheme. Variations of this scheme are still being used today (for example, Park and MacDiarmid, 1975). More recent, and traditionally, European workers, have not been satisfied with the emphasis that Lindgren's scheme had on the importance of igneous rocks in ore genesis. Stanton (1972), for example, considers mineral deposits in terms of their associated rocks. The increasing move to an emphasis on sedimentary genesis for mineral deposits has generated considerable discussion. The terms stratabound and stratiform are being used in mineral deposit descriptions (Figure 2.1). Stratabound deposits are contained within one lithological layer (which may be sedimentary or igneous). Stratiform deposits have the form of sedimentary strata which can also be sedimentary or igneous in origin. An interesting review of these terms is in Canavan (1973).

All major texts on mineral deposits review the various classification schemes and theories of ore genesis and their development. Bateman (1958), Stanton (1972) and Park and MacDiarmid (1975) all provide interesting and informative reading on these topics.

THE FORMATION OF MINERAL DEPOSITS

Introduction

For centuries, mineral deposits were considered to be formed by exotic or extraordinary processes other than those which form the rocks of the Earth. It was not until the end of the 18th century that investigators began to seriously consider that mineral deposits may have formed in a similar manner to ordinary rocks. A consequence of this approach is that deposits could be considered to be rocks which have a chemical and/or mineralogical composition which may be of some value to our civilisation.

For example, Abraham Werner, the proponent of the Neptunist theory for the origin of rocks, considered that mineral deposits were simply special chemical sediments. This is consistent with his theory that all rocks were the result of precipitation from a primaeval ocean.

Similarly, James Hutton, leader of the Plutonist school, proposed that mineral deposits were formed by intrusion of molten sulfurous material. This is consistent with his theory that igneous rocks were formed from molten rock material.

Obviously, the above two views are extreme. However, it is significant that both workers saw the formation of mineral deposits not as particularly special events but as part of the broad scheme of rock forming processes.

Igneous Mineral Deposits

Introduction The stages in the crystallisation of a magma may be summarised:

1. Magmatic Stage - where there is equilibrium between liquid (silica melt) and crystalline phases.
2. Pegmatitic Stage - where there is equilibrium between liquid, crystalline and gas phases.
3. Pneumatolitic Stage - where there is equilibrium between crystalline and gas phases.
4. Hydrothermal Stage - where there is equilibrium between crystalline, aqueous solution and gas phases.

Note that it is not unusual to symplify the above scheme by combining 2 and 3 into the generalised pegmatitic stage.

Magmatic Deposits Mineral deposits formed during the magmatic stage are of great importance to our civilisation.

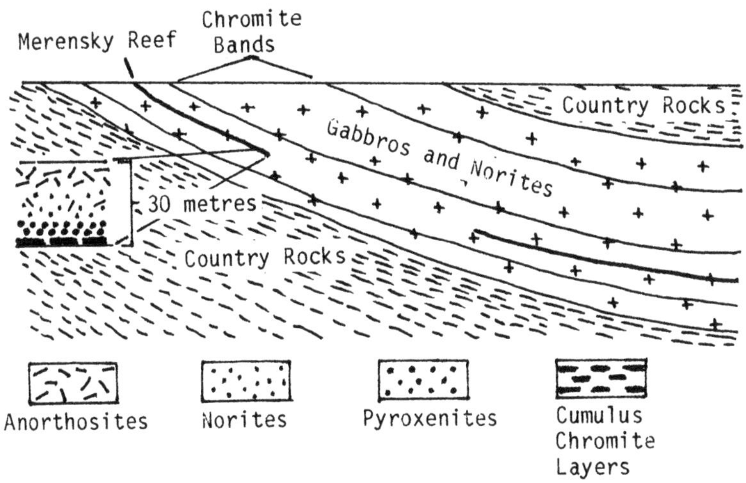

Figure 1.2: Simplified Diagram of the Bushveld Igneous Complex Showing Details of the Merensky Reef.

One example is the Merensky Reef, part of the Bushveld Complex in the Transvaal of Africa. This complex is 460 by 240 kilometres in areal extent, 8000 metres thick and has a lopolithic form. Successive intrusions of granodiorites, diorites, gabbros and ultramafic magmas were differentiated by gravity settling. During this process early crystallising chromite settled to the bottom of most intrusions (Figure 1.2.).

The Merensky Reef is one of these differentiation zones which, in simplistic terms, grades from anorthosite at the top downwards through norite to pyroxenite with chromite bands at different levels. Associated with these chromite bands (which are of economic importance in their own right) are platinoid minerals (such as platinum and paladium) which are concentrated within or adjacent to two of the chromite bands.

The general theory suggests that the early crystallising, high temperature, high density minerals settle down through the magma to form layers of economic importance. The Pallisades Sill in New Jersey, U.S.A., is a well documented example of similar magmatic differentiation.

The Western Australian Kambalda nickel sulfide deposits are considered to have formed in a similar way. Earlier workers (Woodal and Travis, 1969) considered that the ultramafic host rocks are sills. However, Ross and Hopkins (1975) suggest that the host rocks are ocean floor extrusives.

<u>Pegmatitic Deposits</u>. Pegmatites may be simple or complex, acid to basic in composition and are commonly zoned to various degrees. Simple pegmatites are more likely to have been formed by metamorphic processes and are unlikely to be of economic importance.

In contrast, complex pegmatites tend to be igneous in origin. Many of these pegmatites are worked because the large size of the crystals enables easy sorting which results in high purity of the end product. The value of some minerals depends on their absolute size, for example, mica sheets used as electric wire supports in electric toasters. In many other cases, complex pegmatites are important sources of rare minerals. One example is the Mountain Pass deposit in California from which a variety of rare earth elements is recovered, mainly from the mineral bastnaesite which is a rare earth carbonate.

<u>Hydrothermal Deposits</u>. There is little doubt that many mineral deposits resulted from the deposition

of the ore minerals from hot, aqueous solutions. It is the origin of these solutions which has been controversial for centuries. During the first half of this century there was a bias towards a magmatic source. Now, however, a more balanced view suggests that the origin of such solutions may be magmatic metamorphic, or even slightly heated circulating meteoric waters.

A substage of the hydrothermal stage of magmatic crystallisation is characterised by the presence of copper, lead and zinc. These metals are considered to be transported in the form of halogens. When such solutions reach a suitable physical - chemical environment the metals are deposited, usually as sulfides.

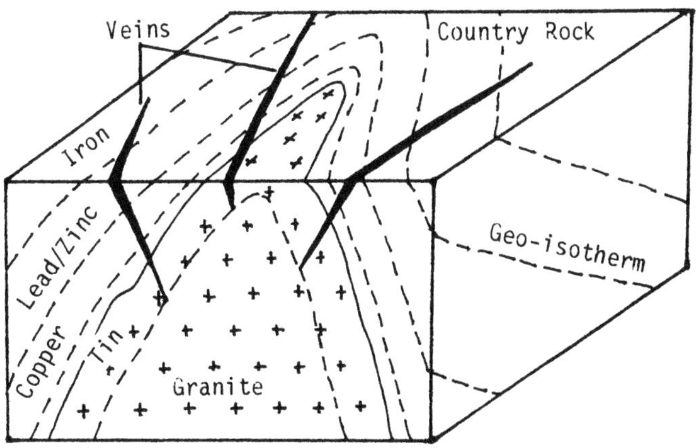

Figure 1.3: Mineral Zoning Associated with the Cornwall Granites.
(After Hosking, 1962).

One region in which such mineralisation occurs is Cornwall in south-west England. Veins of tin, copper, lead and zinc have an obvious genetic relationship to the several granite intrusions of the area. The interesting feature about these deposits is that the tin mineralisation occurs in the granite or in the immediately surrounding contact metamorphic rocks. Copper mineralisation is found further out from the granite and lead-zinc veins further still (with non-economic iron rich veins succeeding these). This pattern of distribution is explained by the temperatures of crystallisation of the various ore minerals. This area is thus a classic example of zoning of mineralisation (Figure 1.3). The area is also a metallogenic province because of the abundance of one type of mineralisation. The mineralisation and the associated granites were emplaced at one period of time known as a metallogenic epoch.

Recently, more interest has been shown in hydrothermal fluids which are formed by metamorphic processes. Sediments of marine origin contain connate (or formation) water rich in halogens, particularly chlorine. When such halogen-rich waters are heated by metamorphic processes - mainly regional metamorphism - they are able to dissolve significant quantities of the trace metals which are present in all rocks. The migration and deposition of the metal-rich solutions can then be considered in the same way as if they were of magmatic origin. There is a growing tendency to compare such processes with those of petroleum formation and deposition, that is, a source rock, migration of the fluid and deposition in a suitable trap.

Late-stage hydrothermal activity can sufficiently alter earlier formed rocks to produce a mineral deposit. A classic example is the China-clay district of Cornwall, where many of the granites have been altered to produce very large deposits of clay.

The same origin is suggested for the "wall-rock alteration" (usually bleaching) associated with many mineral deposits. Porphyry copper deposits, for example, are associated with granodiorite to quartz monzonite plutons of Phanerozoic age. The source of the metals found in these deposits is considered to be both from the plutons and from the surrounding country rocks. In the latter case, the metals in the country rocks are leached by circulating hydrothermal fluids associated with the intrusions; an envelope of bleached rocks is a by-product of this process. Figure 1.4 illustrates the relationship between alteration and mineralisation. There are three main zones of alteration:

1. Potassic zone which is developed in the intrusive body and contains potassium feldspar and biotite
2. Phyllic zone which is developed in the contact zone between the intrusion and the country rocks and contains quartz, sericite and pyrite

3. Propylitic zone which is developed in the country rocks and consists of albite, chlorite, epidote and calcite.

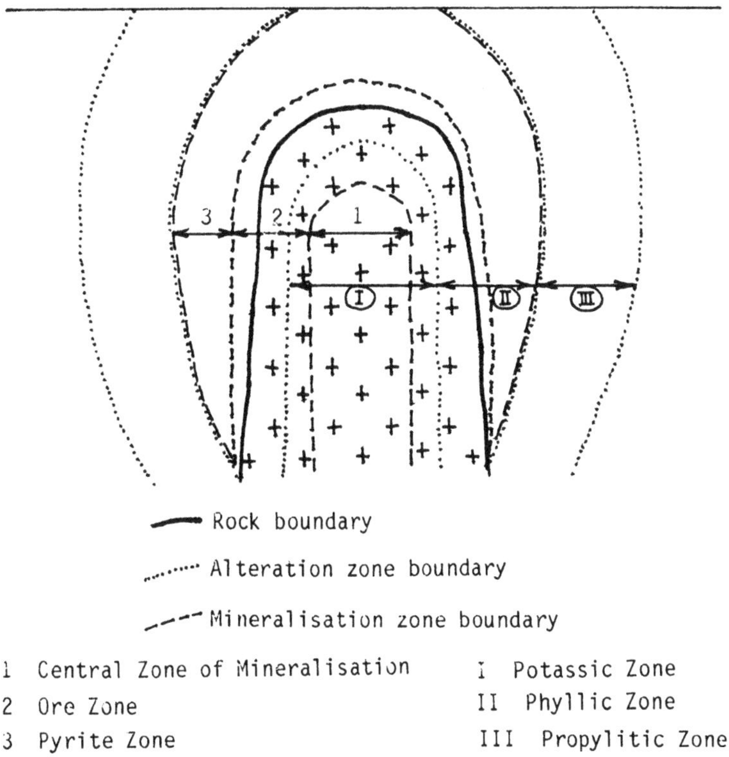

Figure 1.4: Ideal Porphyry Copper Deposit Showing Zones of Mineralisation and Alteration.

Related to the above zones are three zones of mineralisation:

1. Central zone which is developed in the intrusion (and in the potassic zone). There is generally only weak copper mineralisation.
2. Ore zone or shell which is developed at the boundary of the potassic and phyllic zones. This zone contains a large proportion of chalcopyrite compared to pyrite.

3. Pyrite zone or shell which is developed in the phyllic zone. This zone contains a small proportion of chalcopyrite compared to pyrite.

Porphyry copper deposits are thus epigenetic, hydrothermal and zoned deposits. For further reading see Titley (1982).

Traditionally (for example, by Lindgren) hydrothermal fluids and their associated deposits have been classified as hypothermal (high temperature and pressure), mesothermal (medium temperature and pressure) and epithermal (low temperature and pressure).

Sedimentary Mineral Deposits

Mineral deposits formed by sedimentary processes are also of great economic importance. However, there is a problem of definitely attributing the origin of deposits to sedimentary processes with increasing geological age. This is because other geological processes such as metamorphism and tectonism tend to mask the original character of the deposit.

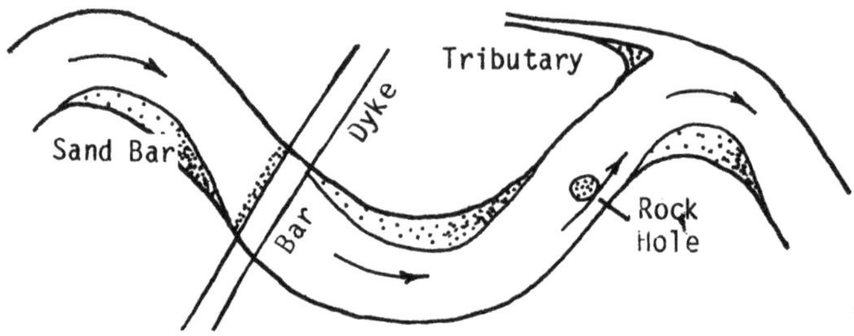

Figure 1.5: Possible Locations of High Density Mineral Deposits in a Stream.

Obviously, mineral deposits which one can see forming can be classified with a reasonable degree of surity. One such example is an alluvial (or placer) gold deposit. In any stream in which there are materials of different densities, if all else is equal, the minerals of higher density will tend to drop to the bottom if the stream velocity decreases sufficiently. Such locations in a stream may be at the confluence of tributaries, rock bars, rock holes and the inside of meanders (Figure 1.5). A Precambrian placer is the gold-bearing quartz conglomerate of the Witwatersrand of South Africa.

Another sedimentary mineral deposit is the "heavy" mineral sand deposit in beach sands. If a coastal region has a hinterland of igneous and/or metamorphic rocks then the coast will be supplied with heavy minerals such as ilmenite, rutile, magnetite, zircon, monazite and so on, as well as tin, gold, platinum and diamonds. These minerals will survive the transportation to the coast because they are relatively resistant to physical and chemical weathering. Once at the coast, these minerals will form part of the beach sands and be distributed along the coast by longshore drift. Storm waves tend to winnow the less dense minerals from the foot of the back - beach dunes leaving a relatively enriched zone of heavy minerals. These include rutile, ilmenite and monazite, and in favourable areas, gold and platinum.

There is an area of overlap between the strictly sedimentary and strictly magmatic origins for some deposits. The Kuroko, or Black Ore deposits of Japan, for example, are considered to have formed as the result of submarine volcanoes expelling metal bearing fluids into the oceanic waters. Subsequent chemical reaction and cooling results in a sedimentary deposit on the sea floor.

Clastic sedimentation may form sand and gravel deposits vital to the construction industry. Chemical and organic sedimentation form limestones used in the construction industry both as a building stone and in the manufacture of cement. Evaporite sedimentary mineral deposits of halite and gypsum are also of great importance. Precambrian banded iron formations are also considered to have formed by cyclic chemical sedimentation.

Residual Deposits

Residual deposits are formed when the unwanted minerals are removed during weathering resulting in a relative concentration of the mineral of value.

Lateritisation of a peneplain can result in deposits of many different minerals such as hematite-limonite, bauxite and lateritic nickel. The type of deposit depends on the nature of the original rock type. The process requires that:

1. The required minerals are essentially insoluble
2. The unwanted minerals are soluble
3. The topographic relief is low so that chemical weathering predominates and there is little chance of the desired minerals

being washed away.
4. The area is tectonically stable so that the process can continue for a long time and form a significant deposit.
5. The climatic conditions are favourable - preferably tropical monsoonal to high rainfall Mediterranean.

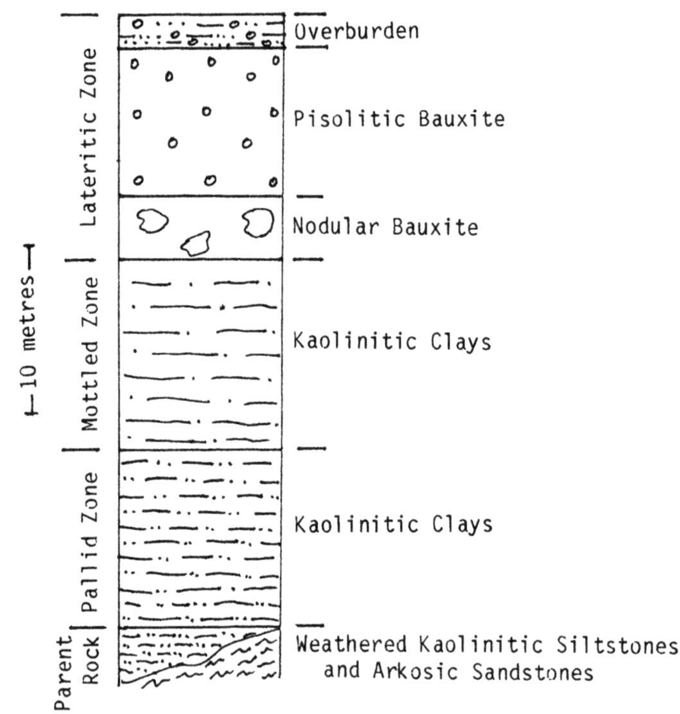

Figure 1.6: Generalised Section Through the Wiepa, Australia, Blanket Bauxite Deposit. (After Evans, 1965).

This process can result in the pisolitic blanket type of bauxite deposits such as the Weipa bauxite deposit in Queensland, Australia (Figure 1.6.) and the lateritic nickel deposits found in New Caledonia.

Secondary Enrichment

Secondary enrichment of a mineral deposit may result in very rich ore or even upgrade an uneconomic deposit to an ore deposit. This process involves dissolving and transporting downwards some of the ore materials and redepositing them at or near the water table.

Figure 1.7 illustrates a copper sulfide vein as an example. The downward percolating meteoric water, which contains dissolved atmospheric oxygen, will oxidise the sulfide ore minerals above the water table. In the case of pyrite (FeS_2), the iron is released to form insoluble iron oxides which remain at the surface as a "gossan". The sulfur

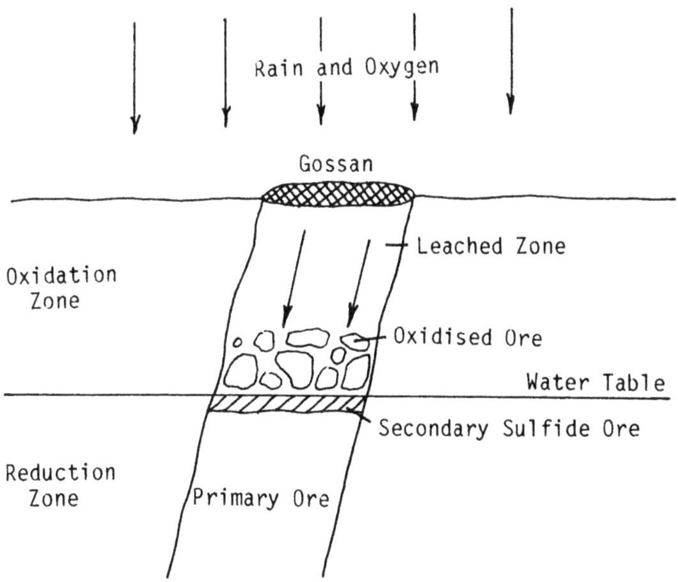

Figure 1.7: Secondary Enrichment of a Mineral Deposit.

combines with the weak meteoric acids to form weak sulfuric acid. This further dissolves the minerals in the zone of oxidation. The metal-rich solutions migrate down to the water table where there is a reducing environment and secondary copper sulfides, such as bornite and covellite, are deposited. If the solutions are near to saturation in the zone of oxidation then copper oxides and carbonates may be deposited above the water table.

Again, as in the case of residual deposits, to

produce a deposit of significant size, climatic, topographical and geological conditions must be favourable. For example, the Flinders Ranges of South Australia have been subject to several episodes of uplift. This has not been favourable for to the formation of large secondary deposits as erosion caused the water table to fall too rapidly resulting in the oxidation of any secondary sulfide deposits. Also, the majority of the country rock is carbonate rich. These rocks neutralised the downward percolating acidic copper-bearing solutions and their metal content was deposited as carbonates (mainly malachite) above the water table. Thus the hopes of the miners, who were mainly of Cornish origin and who expected rich secondary sulfide deposits after encountering such rich ore in the oxidised zone, were in vain as they had mined all the secondary copper as carbonates above the water table.

Weathering

Finally, weathering must be recognised as an important process in the formation of mineral deposits. The simplest examples are the weathering of rocks to produce clays for pottery or brick-making. Weathering can also change the economics of a mineral deposit. For example, it is very costly to mine and treat kimberlite to extract diamonds. However, if the kimberlite has been sufficiently weathered to form clays then the recovery of diamonds is a much simpler and less costly exercise.

CONCEPTUAL MODELS OF MINERAL DEPOSIT FORMATION

One of the more fascinating aspects of economic geology is the attempt to develop a universal theory of mineral deposit formation. Any such theory must consider the source of the material, its transport and subsequent deposition. The transporting medium is normally accidental to the process, the mineral material simply "hitching a ride". For example, hydrothermal fluids may originate in a metamorphic environment, move through a metal-bearing rock and dissolve the metal, transport the metal to another environment, where the conditions are such that the metal is deposited, and continue on their way.

Mineral deposition, or ore genesis, has been considered by various workers in the light of:

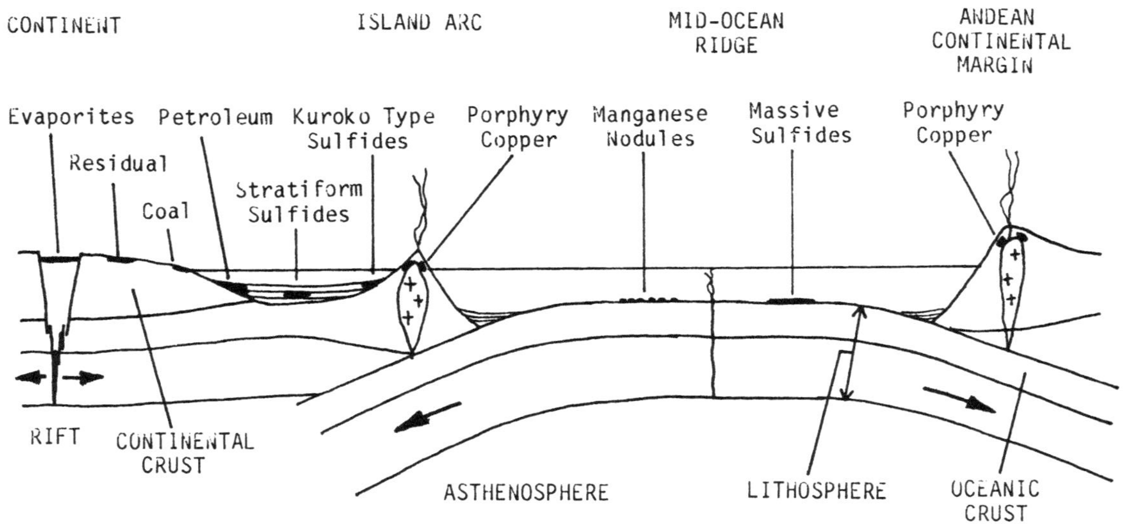

Figure 1.8: Suggested Relationships Between Some Mineral Deposits and the Main Elements of the Plate Tectonic Theory.

1. Tectonism in general
2. Continental drift
3. Plate tectonics
4. Crustal evolution
5. Atmospheric evolution
6. Geochemistry

The origin of many sedimentary mineral deposits is related to the latitude of the depositional environment. For example, phosphorite deposits form in relatively low latitudes. However, subsequent continental drift has distributed these deposits over a wide range of present day latitudes.

Figure 1.8 attempts to relate several different types of mineral deposits to the major elements of the plate tectonic theory.

These, and similar concepts, are treated in the comprehensive list of papers in Wright (1977). Not only are metallic mineral deposits represented in this collection of papers, but so also are coal, petroleum and other types of deposits.

REFERENCES

Aus. I. M. M., 1981. Reporting of Ore Reserves. Report of the Joint Committee of the Aus. I. M. M. and AMIC. Aus. I. M. M. Bull. No 452: 9-11.

Bateman, A. M., 1950. Economic Mineral Deposits (2nd Ed.), Wiley and Sons, New York.

Canavan, F., 1973. Notes on the Terms "Stratiform", "Stratabound" and "Stratigraphic Control" as Applied to Mineral Deposits. J. Geol. Soc. Aust., **19**:543-546.

Evans, H. J., 1965. Bauxite Deposits of Weipa, in McAndrew, J. (Ed.), 1965. Geology of Australian Ore Deposits (2nd Ed.). Aus. I. M. M., Melbourne.

Hosking, K. F. G., 1962. The Relationship Between the Primary Mineralization and the Structure of South-West England, in Coe, K. (Ed.), 1962. Some Aspects of the Variscan Fold Belt. University Press, Manchester.

King, H. F., McMahon, D. W. and Bujtor, G. J., 1982. A Guide to the Understanding of Ore Reserve Estimation. Aus. I. M. M., Melbourne.

McDivett, J. F. and Manners, G., 1974. Minerals and Men (Revised Ed.), John Hopkins University Press, Baltimore.

Park, C. F. and MacDiarmid, R. A., 1975. Ore Deposits (3rd Ed.). Freeman, San Francisco.

Ross, J. R. and Hopkins, G. M. F., 1975. Kambalda Nickel Sulphide Deposits, in Knight, C. J. (Ed.), 1975. Economic Geology of Australia and Papua-New Guinea. Mono. Series No. 5, Part 1, Metals, Aus. I. M. M., Melbourne.

Skinner, B. J., 1976. Earth Resources (2nd Ed.). Prentice-Hall, Inc., Englwood Cliffs.

Stanton, R. L., 1972. Ore Petrology. McGraw-Hill, New York.

Sykes, T., 1978. The Money Miners. Wildcat Press, Sydney.

Titley, S. R. (Ed.), 1982. Advances in Geology of Porphyry Copper Deposits. University of Arizona Press, Tucson.

Woodal, R. and Travis, G. A., 1969. The Kambalda Nickel Deposits, Western Australia. Proc. 9th. Common. Min. Metall. Congress, London.

Wright, J. B. (Ed.), 1977. Mineral Deposits, Continental Drift and Plate Tectonics. Benchmark Papers in Geology, **44**. Dowden, Hutchinson and Ross, Inc, Stroudsburg.

FURTHER READING

TEXTS

Evans, A. M., 1980. An Introduction to Ore Deposits. Blackwell Scientific, Oxford.

Wolf, K. H. (Ed), 1976. Handbook of Strata - Bound and Stratiform Ore Deposits (7 vols). Elsevier Scientific Publishing Co., Amsterdam.

Mineral Deposits

JOURNALS

Economic Geology
Mineralium Deposita
Mining Annual Review
Mining Magazine
Mining Journal
Proceedings of the Australasian Institute of Mining and Metallurgy
Transactions (Section B) of the Institution of Mining and Metallurgy

Chapter Two

COAL

INTRODUCTION

Coal is a carbonaceous rock formed by the accumulation, alteration and preservation of predominantly land plant matter. It is one of the fossil fuels. Coal is a combustible, opaque (except in thin sections), non-crystalline solid which varies from light brown to black. Lustre varies from dull to brilliant, density from 1.0 to 1.8 g cm^{-3} and hardness from 0.5 to 2.5. Coal is brittle and has a hackly fracture.

The main chemical constituents are carbon, hydrogen and oxygen with lesser amounts of nitrogen and sulfur. The ash or mineral matter can be quite variable. There is a general inverse relationship between oxygen and, to a lesser extent, hydrogen content to the carbon content of a coal.

Coal is a large complex molecule consisting mainly of carbon rings. Carbon atoms may be replaced by oxygen, nitrogen or sulfur atoms. It is this sulfur, as well as the sulfur in associated discrete sulfur bearing minerals such as pyrite and gypsum, that contributes to many environmental problems such as "acid rain".

There are few texts which deal with coal on a geological basis only. However, a number of texts consider coal in terms useful to geologists. These include Pettijohn (1975), Speight (1983), Van Krevelen (1961), Tissot and Welte (1978) and Williamson (1967).

OCCURRENCE

Coal is relatively rare but widely distributed, both in time and space, although it did not become common until the development of woody plants in the

Devonian. The majority of older coals are sapropelic. That is, they are composed of amorphous organic matter such as spores and planktonic algae. The first extensive coals are of Carboniferous age (hence the term). Such coals are generally restricted to the northern hemisphere. The later Permian coals generally occur in the southern hemisphere (with the notable exception of Indian coals).

In many coal basins there are several coal seams or beds interbedded with shales and sandstones. Such a sequence is referred to as coal measures. Often, successive coal seams and their associated sedimentary rocks show a cyclic arrangement. A series of beds deposited during one of these cycles is called a cyclothem (Figure 2.1).

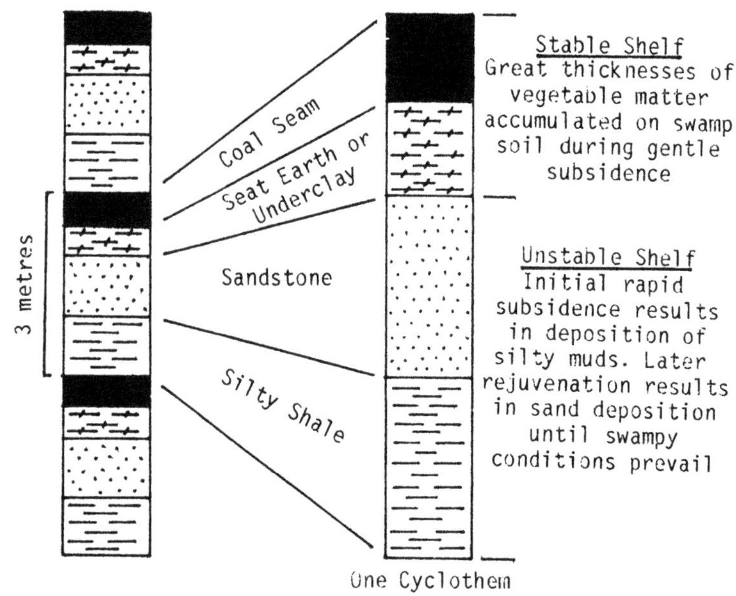

Figure 2.1: A Simplified Series of Cyclothems with an Interpretation of the Lithologies. (After Krumbein and Sloss, 1963; and Read and Watson, 1968).

The cyclicity of the stable and unstable shelf conditions of sedimentation as indicated in Figure 2.1 may be explained in terms of continental drift. During the separation of continental masses, basins of internal drainage are formed by rifting and the downthrown blocks rejuvenate the drainage systems.

With the passage of time, these basins become filled with sediment and swampy conditions develop with the associated formation of peat. As the continental masses continue to move apart further rifting occurs with drainage rejuvenation and burial of the peat. Such cycles may continue until continental separation is complete and marine water enters the basins. The stratigraphy of the northern part of Bass Strait, Australia, well illustrates this cyclicity.

FORMATION

Accumulation

In Situ Theory. Most coals are considered to have formed from plant matter which accumulated at the place of growth. The coals that result are called autochthonous. The evidence supporting this theory is:

1. The plant debris is not sorted
2. The seat earth contains roots penetrating from the overlying coal
3. There is a general lack of clastic material indicating still water conditions
4. There is abundant evidence from present fresh water swamps to support the possibility of great thicknesses of peat. Examples are the peat mosses of the muskeg of Canada and the peats of the coastal regions of the tropics. One such fresh water area in Sumatra is approximately 700 square kilometres in areal extent and contains 10 metres of peat and brown coal of Miocene age
5. Large coal fields suggest that accumulation was at or near sea level so that slight subsidence would cause large scale flooding

Drift Theory. Some coals are considered to have formed from material which has been transported to the present location to form allochthonous (or transported) coals. Supporting evidence is:

1. Not all coals are underlain by seat earths
2. Sedimentary structures in the coals suggest that the material has been transported, for example, rafts of humus and coal-forming plant matter
3. The fine-grained nature of some coals (for example, cannel coals) and their high ash content (indicative of fine clastic mater-

ial) suggests sorting by normal sedimentological processes

4. The fact that not all coals are associated with cyclothems

Both methods of accumulation do occur and the origin of each coal deposit should not be determined by dogma but by careful observation and deduction.

Coalification

The process of conversion of woody and peaty residues to coal is called coalification. For successful formation of thick deposits large thicknesses of plant material must be allowed to accumulate and, once accumulated, must be preserved from decay.

In tropical climates the rapid growth of plants more than compensates for the rapid decomposition by bacteria which are very active in such environments. In arctic climates there is slow accumulation of plant material (such as in the muskegs of Canada). However, there is little decomposition by bacteria which are not very active in such an environment.

Slow subsidence of the basin of deposition or the encroachment of swamps into lakes will allow thick deposits to accumulate. As a general rule approximately 10 metres thickness of plant material will produce one metre of coal. Once a suitable thickness of plant material has accumulated it will be preserved by rapid burial by sediment.

The coalification process itself starts as soon as plant material accumulates under water in an aerobic environment. Initially, diagenesis or biochemical coalification occurs. Partial decomposition of the plant debris is carried out by bacterial, fungal and microbial activity, until accumulation of waste products kills the decomposers. The material thus formed is peat and this part of the coalification process results in the relative increase of carbon in the coalified matter as compared to that in the original cellulose.

Physical and chemical coalification then results in the change of peat to the various ranks of coal. This change is dependent on the pressure and temperature conditions during the subsequent burial of the coal. Temperature is considered to be the more important factor particularly in relation to the length of time that the coal has been buried.

CONSTITUENTS

Macerals

A coal maceral is a single organic unit such as a single fragment of plant debris. Macerals have also been called the "coal minerals". Maceral names always end in "-inite". The various macerals are described below:

1. Vitrinite consists of collinite, a structureless jellified plant residue, and telinite, a translucent golden gel that retains some cell structure. (All colours given are those observed under transmitted light).
2. Exinite consists of the smaller and more resistant plant debris such as:

   (a) alginite - of algal derivation
   (b) sporinite - spore cases
   (c) resinite - composed of plant resin
   (d) cutinite - macerated fragments of cuticle

   Sporinite is yellow and transparent and is composed of both small (microspores) and large (megaspores) spore bodies. Resinite occurs as small isolated bodies of a reddish translucent nature.
3. Fusinite is the opaque-walled carbonised cell structure - "mineral charcoal".
4. Micrinite is an opaque residue.
5. Sclerotinite is formed from fungi.

Anthraxylon and attritus are alternative coal constituent definitions (to those of 1 to 4 inclusive) adopted by some North American coal petrologists:

1. Anthraxylon is translucent material derived from the woody parts of trees.
2. Attritus is opaque and translucent macerated and degraded plant debris. This term is applied to all material not identified as anthraxylon.

Lithotypes

Coal lithotypes are easiest applied to bituminous (or humic or banded) coals above lignite in rank. Any one specimen of these coals usually consists of one or more of the coal lithotypes, usually in bands. The bright bands are composed of the major structural parts of plants, especially wood - the

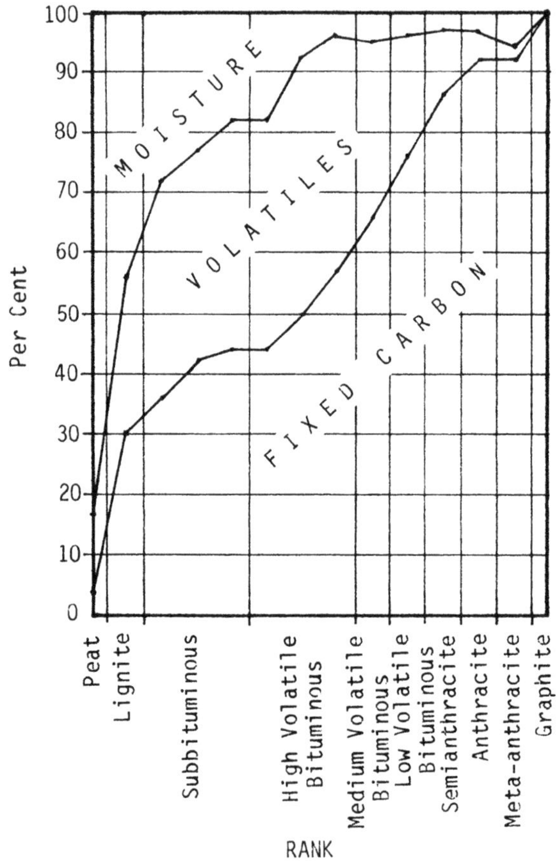

Figure 2.2: The Coal Series.

vitrain and clarain lithotypes. The dull bands composed of plant debris such as leaves, spores and cuticle - the fusain and durain lithotypes. Lithotype names end in "-ain". They are described below:

1. Vitrain is composed primarily of vitrinite. It is brilliant, vitreous, jetlike with a conchoidal fracture and is clean to the

touch. In North American terminology vitrain is composed primarily of anthraxylon. It is usually structureless.

2. Clarain is composed primarily of vitrinite and some exinite and other macerals. It is laminated, has smooth fracture, a shiny lustre and a silky appearance due to the laminations. Alternatively, clarain is composed of translucent attritus and thin shreds of anthraxylon.
3. Durain is dominantly micrinite (opaque detritus) with exinite. It is a dull coal with an earthy lustre, black to lead grey, close texture generally with no internal stratification.
4. Fusain ("mineral charcoal") is composed of fusinite and is similar to ordinary charcoal. It is unmineralised and very friable and porous. Often the pores are filled with calcite or quartz. In thin sections, fusain is opaque and highly cellular.

CLASSIFICATION

Rank is based on the stage that chemical alteration has reached - the degree of coal metamorphism or lithification. Figure 2.2 shows the relationship between rank and the proportions of the main chemical constituents. This figure applies to the "normal" coals - that is, those made up predominantly of wood and bark, the humic coals of the Coal Series.

The type of coal is determined by the nature of the plant residue. For example, the normal coals of the coal series are banded (Figure 2.3). The upper durain band in the figure shows orange exinite as sporinite (both micro- and mega-spores) in black micrinite. The vitrain band is composed of red collinite (structureless vitrinite). The lower durain band is composed of orange exinite in structureless micrinite.

In contrast cannel and boghead coals are clean, compact, blocky coals of massive structure and uniform, fine grained texture. These coals are normally grey to black, have a greasy lustre and conchoidal fracture.

Cannel coal has no bedding and is composed of microdebris rich in inflammable hydrocarbons such as spores, resins, woody fragments and opaque attritus. The name is derived from "candle" as splinters of this coal burn like a candle. Cannel coal is intermediate between bituminous and boghead coal (Figure 2.4).

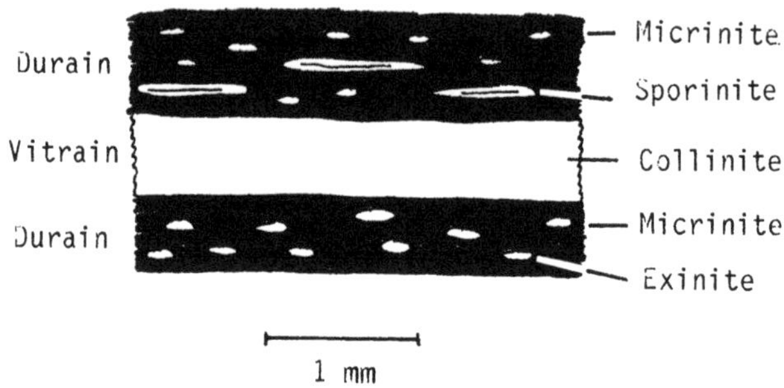

Figure 2.3: A Typical Banded Coal.

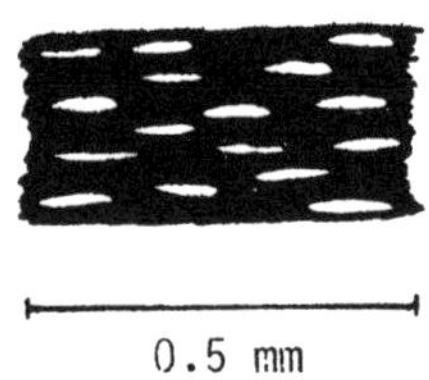

Figure 2.4: Cannel Coal.
Elongate Blebs of Red Vitrinite and Orange Exinite in Black Micrinite.

Boghead coal is rich in algal remains and has a high ash content. Bone coal is a very impure, high ash-content coal. There is a continuous series from the humic coals of the coal series through cannel coal and boghead coal to sapropelites. There is another series from the humic coals through bone coal to ordinary shales (Figure 2.5).

Humic Coals
↙ Bone Coal ↙ Black Shale ↙ Shales
↘ Cannel Coal ↘ Boghead Coal ↘ Sapropelites

Figure 2.5: The Relationship of Humic Coals to Shales and Sapropelites. (After Pettijohn, 1975).

REFERENCES

Krumbein, W. S. and Sloss, L. L., 1963. Stratigraphy and Sedimentation (2nd Ed.). Freeman and Co., San Francisco.

Pettijohn, F. J., 1975. Sedimentary Rocks (3rd Ed.). Harper and Row, New York.

Read, H. H. and Watson, J., 1962. Introduction to Geology (2nd Ed.). Macmillan, London.

Speight, J. G., 1983. The Chemistry and Technology of Coal (Chemical Industries, v 12). Dekker, New York.

Tissot, B. P. and Welte, D. H., 1978. Petroleum Formation and Occurrence. Springer-Verlag, Berlin.

Van Krevelen, D. W., 1961. Coal: Typology - Chemistry - Physics - Constitution. Elsevier Publishing Co., Amsterdam.

Williamson, J. A., 1967. Coal Mining Geology. Oxford University Press, London.

Chapter Three

PETROLEUM

INTRODUCTION

There are a number of very good texts on petroleum geology and geochemistry. Reference is made to many of these in this chapter. Levorsen (1967) is probably the best known of the petroleum texts. A more recent work is that of Giuliano (1981), a very readable introduction to the petroleum field, covering all aspects from geology and exploration to production and political considerations.

Petroleum includes all naturally occurring hydrocarbons whether gaseous, liquid or solid. A hydrocarbon is any compound consisting only of hydrogen and carbon.

Figure 3.1 illustrates the three main types of petroleum based on their molecular structure. These three types are:

1. Paraffin rich petroleums which have the general composition C_nH_{2n+2}, a straight or branched chain structure and range from methane to paraffin wax. Everyday petroleum produced from the paraffins includes natural gas, liquefied petroleum gas (LPG), petrol and kerosene. These are light crude oils.
2. Naphtha rich petroleums which have the general composition C_nH_{2n} , a carbon ring structure and range from heavy fuel oils to asphalt residue. These are relatively rare crude oils.
3. Aromatic hydrocarbon rich petroleums which have the general formula C_nH_{2n-6}, contain at least one benzene ring and include benzene and toluene. These are usually heavy to very heavy crude oils.

All crude oils contain at least some of each of these three types (Figure 3.2). The proportions of the types depend on the original material from which the petroleum was formed and the subsequent modification of that petroleum.

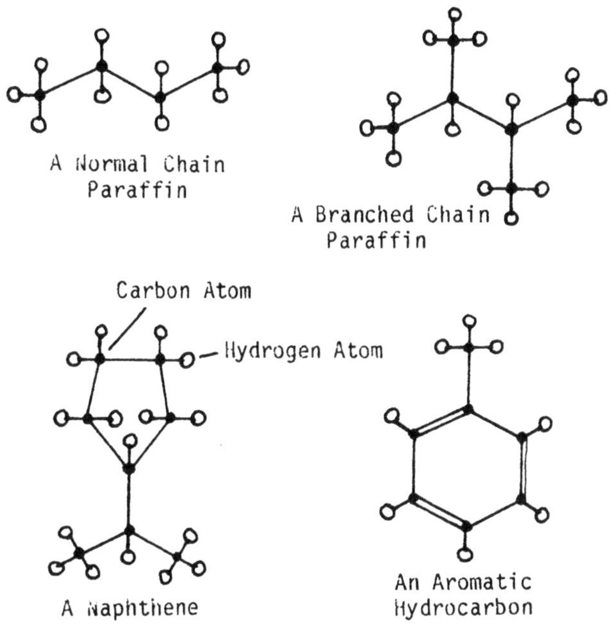

Figure 3.1: The Molecular Structures of the Three Main Types of Petroleum.

FORMATION

Organic Origin

There is a considerable body of evidence to support the theory of an organic origin for petroleum. However, there are several workers who consider that some methane may still be being produced in the Earth's mantle and that this may be a non-organic source.

Evidence for an organic origin is detailed below:

1. Petroleum contains several compounds derived from chlorophyll, which has been in existence on the Earth since the first algal flora about three billion years ago.

Chlorophyll will hydrolyse to form phytol, which will form, by oxidation or reduction, the hydrocarbons pristane or phytane respectively, both common in crude oils.

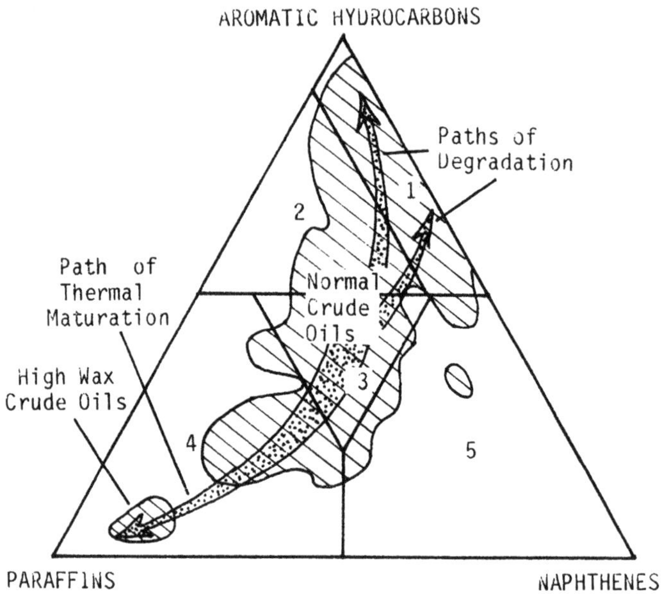

1 - Aromatic heavy degraded oils
2 - Aromatic intermediate oils
3 - Paraffinic - Naphthenic oils
4 - Paraffinic oils
5 - Naphthenic oils

Figure 3.2: Composition, Classification and Evolutionary Paths of Crude Oils. (After Tissot and Welte, 1978).

2. Zooplankton concentrate in their fatty parts the pristane from phytoplankton. Upon the death of the zooplankton the pristane is incorporated into marine sediments.
3. Carotene related hydrocarbons exist which suggest a vegetable origin.

However, not all hydrocarbons formed in living things. For example, none of the light hydrocarbons in the ethane-nonane range have been identified in significant amounts in living things or recent sediments.

It appears that the origin of petroleum is a dual process with some hydrocarbons being formed in living things directly, and other hydrocarbons being formed by the reduction of other organic matter during sediment diagenesis. In the latter case, the role of kerogen is of great importance. Nearly all of the gasoline (petrol) range hydrocarbons and half or more of the heavier hydrocarbons form by the second process, with the rest of the petroleum forming by the first.

Petroleum generally forms in marine environments because they are more reducing than most terrestrial environments. The obvious exceptions are environments which produce coals. In reducing environments, new hydrocarbons will form from organic matter and old hydrocarbons will be preserved. However, the Uinta Basin oil shales in the U.S.A. and the petroleum deposits of the Cooper Basin in Australia are both terrestrial deposits. Although petroleum generally forms in marine environments, the source of the organic material may be of terrestrial origin. Marine derived petroleum tends to have short carbon chain lengths and terrestrial derived petroleum tends to have long chain lengths.

Kerogen

Kerogen is the precursor of most oil and gas. It is the insoluble organic matter in sedimentary rocks and can have its origin in sapropelites or peats. Sapropelites are muds which form in reducing conditions and contain the decomposition products of organic material such as spores and planktonic algae. Diagenesis of sapropelites results in the formation of boghead coals and oil shales.

Hobson and Tiratsoo (1981) describe three types of kerogen:

1. Type I with an algal organic source. This type forms oil shales and boghead coals and has a high potential for producing petroleum rich in paraffins with some naphthenes and aromatics
2. Type II with a sapropelic source mainly of planktonic matter. This is a typical source rock and produces naphtha and aromatic rich petroleum
3. Type III with a terrestrial plant (humic) origin. This type normally produces dry gas only

Petroleum Generation

Thermodynamic considerations suggest that given the right organic material, it would take a million years to form a petroleum accumulation, assuming a minimum overburden of about 600m. Because of the Earth's geothermal gradient, the thicker the overburden, the faster petroleum will be generated. However, as depth increases, porosity and permeability decreases so that it is less likely that any petroleum formed will be able to migrate to suitable traps.

In addition, the deeper the sediment, the lighter the fraction of petroleum produced. This is because the deeper the sediment the higher the temperature which in turn determines the degree to which natural cracking occurs. The relationship between the type of kerogen, vitrinite reflectance and the type of petroleum produced (oil or gas) is given in Figure 3.3.

The relationship between kerogen and coal maceral types is clear. The early evolution of the organic material (during diagenesis) results in only a slight decrease in H/C ratios but a marked decrease in O/C ratios. This shows that little hydrocarbon material was formed from the degradation of the kerogen.

The second stage of the evolution, catagenesis, results in a large decrease in the H/C ratio. It is during this stage that the kerogen is degraded to form hydrocarbons. Oil is formed first. Then, with increasing temperature, wet gas is produced. Finally, the kerogen is subjected to metagenesis when the temperatures are so high as to allow the formation of dry gas only.

Vitrinite reflectance (expressed as Ro in per cent) is a measure of petroleum maturation and closely parallels coal rank. It is used to determine the type of petroleum likely to be found in a sediment. In the example of Figure 3.3, a sediment with Ro ranging from 0.5 to about 1.35 (corresponding to bituminous rank) is likely to produce oil. If the original organic material was coaly then gas (mostly dry) is produced in the range of 1.0 to 4.8 (which corresponds to middle bituminous to middle anthracite rank).

The formation of petroleum is an inefficient process. The total amount of organic matter in the fine grained sediments of the world has been estimated to be about 3000 trillion tonnes with associated hydrocarbons of 60 trillion tonnes with only 0.6 trillion tonnes in reservoirs (Hunt,

1968).

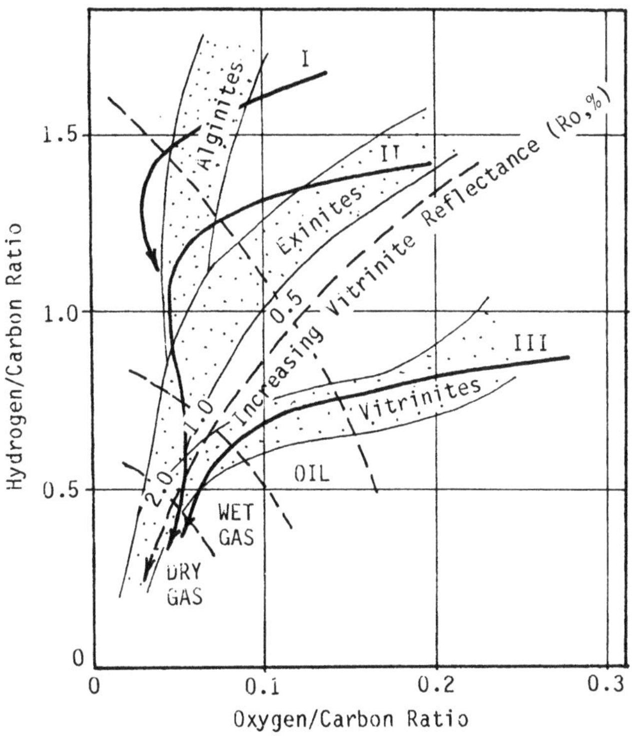

Figure 3.3: Evolutionary Paths of Kerogen Types I, II and III and Coal Maceral Groups with Vitrinite Reflectance Trends and Type of Petroleum Produced.
(After Van Krevelen, 1961; and Tissot and Welte, 1978)

MIGRATION

Organic matter accumulates with fine grained mineral particles. If organic matter is deposited with coarse grained sediments it is flushed out by circulating water and destroyed. Fine grained organic-rich sediments are called sapropelites and constitute the principal source of petroleum.

When the source sediment is buried, the contained water is expelled during compaction. The amount of water is quite considerable because clays have a porosity of about 50% but shales have a porosity of about 3%. The expelled water generally contains light hydrocarbons in the gas to petrol range in solution. Their concentration, 10 - 100 ppm, is sufficient to account for the petroleum in most sedimentary basins. Heavier hydrocarbons cannot be carried in this way and there are several alternative hypotheses to explain their migration.

The direction of movement of the expelled fluids may be in any direction although it is most commonly in a vertical direction. This primary migration, as it is called, continues until the expelled fluids reach a coarse grained rock. Primary migration is thus over relatively short distances.

If the coarse grained rock is also permeable, then the fluid continues to migrate through the rock. A number of forces causes this movement (buoyancy and hydrodynamic pressure are two) to continue until the fluid is trapped and preserved or it reaches the Earth's surface and is destroyed by oxidation. This migration is called secondary migration, and the porous and permeable rock is called a reservoir rock. Impermeable rocks such as shales are called cap rocks. Figure 3.4 illustrates these various terms.

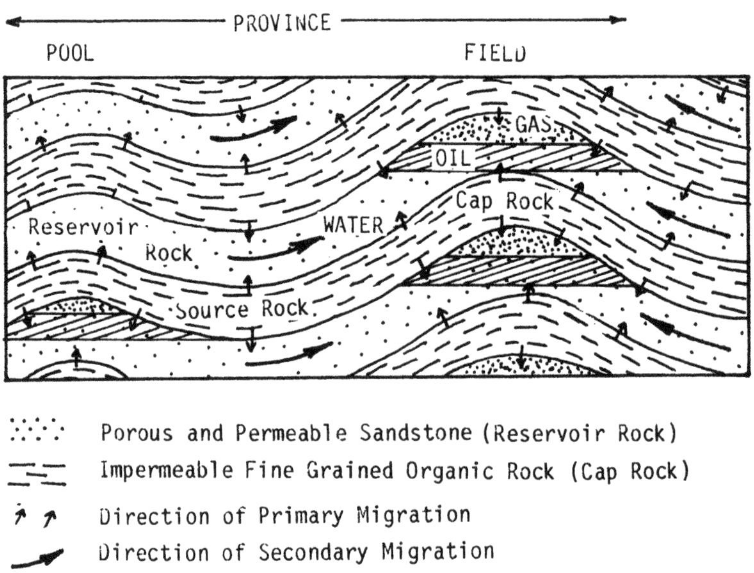

Figure 3.4: Primary and Secondary Migration of Petroleum.

OCCURRENCES OF PETROLEUM

Petroleum and Geologic Time

There is an inverse relationship between the quantity of petroleum and geologic time (Figure 3.5).

Two explanations for this relationship are:

1. The biomass has grown exponentially.
2. The longer a deposit remains in the Earth's crust, the greater the chance of destruction. For example, if a sedimentary sequence is uplifted and eroded, the petroleum is oxidised and rejoins the biogeochemical cycle. Conversely, if the sequence is buried too deeply then only methane and an insoluble carbon-rich residue of very high molecular weight (pyrobitumen) will result.

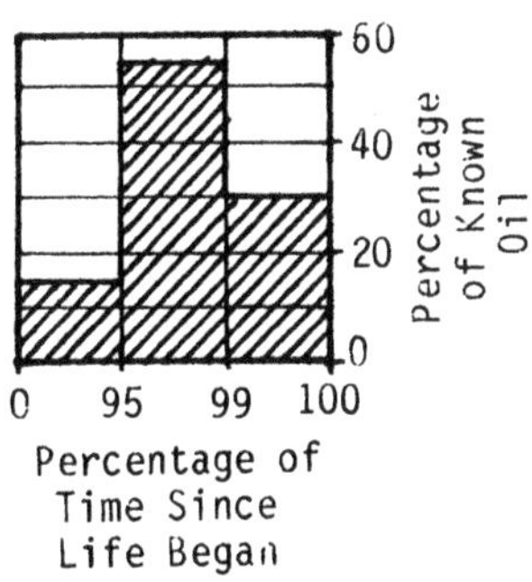

Figure 3.5: Quantity of Petroleum Versus the Age of the Host Sediments. (Data from Hunt, 1968.)

Surface Occurrences

Occurrences can be "live" or "dead". The most common live surface occurrences are seepages, springs and bitumen exudates. These result when a petroleum trap has been breached, usually by faults, and the petroleum reaches the Earth's surface. Some of these may be quite large and of economic importance. For example, Pitch Lake in south-west Trinidad, a large seep of asphalt, gas, water, sand and clay is mined for the petroleum content. The seep constantly replenishes the mined-out area.

Mud volcanoes and mud flows are underlain by

incompetent sediments such as shales and if gas pressure in the underlying reservoir is great enough the gas reaches the surface. The most famous of these is the Baku area on the west side of the Caspian Sea. Spontaneous combustion of the gases resulted in the "everlasting" fires recorded in the Bible.

Surface showings of petroleum can reach extremely large sizes. The largest and most well known is the Athabasca oil sands in Alberta, Canada. Here, Cretaceous sediments were mixed with oil, asphalt and bitumen during deposition. The sediments were subsequently buried but have now been re-exposed. This deposit is considered to be the largest petroleum deposit in the world with an estimated 600 billion barrels of oil in place, half of which is considered to be recoverable.

Another surface solid petroleum occurrence is oil shale. The potential of these deposits periodically receives much attention depending on the current state of supply from the Organisation of Petroleum Exporting Countries (OPEC). Oil shales are widely distributed in time and space. They are organic and bituminous shales, the organic matter being the mineraloid kerogen. If kerogen is heated a small amount of oil is produced. The average yield of an oil shale being about 100-200 litres per tonne. Oil shales have been worked infrequently for most of this century but with the uncertainty of OPEC oil supply and prices oil shales may be the solution to many nations' energy problems.

Subsurface Occurrences

Most wells drilled in sedimentary basins have shows of hydrocarbons. These shows are important clues to the discovery of commercial deposits. They can indicate the presence of a source rock or, more importantly, indicate the presence of a petroleum province or pool.

Commercial deposits are classified as:

1. Pools - the simplest unit. The body of petroleum occurs in an individual reservoir with its own pressure system. Size is not important in this definition
2. Fields are defined where several pools are related to a single geological feature. For example, several overlying anticlinal pools
3. Provinces are regions in which a number of petroleum pools and fields occur in a similar or related geological environment

such as a sedimentary basin (Figure 3.4).

There are many types of petroleum traps, broadly grouped into:

1. Structural - such as the anticlinal traps shown in Figure 3.4
2. Stratigraphic
3. Cementation
4. Hydrologic

Examples of many of these are given in most introductory geological texts.

EXTRACTION

The majority of the world's petroleum is recovered from wells drilled by rotary rigs similar to that illustrated in Chapter 8. In many cases, the pressure of the gas in the reservoir forces the oil to the surface. If gas is intersected, it too flows to the surface. In the cases where there is insufficient pressure to force the oil to the surface then the oil must be pumped out. All these extraction methods are called "primary". Generally less than half of the contained petroleum is recovered using these techniques.

In order to recover more of the contained oil more sophisticated "secondary" techniques are used. These are usually designed to increase the permeability of the reservoir rocks in the vicinity of the well. The techniques include water injection, fracturing and gas repressuring as described in Chapter 10.

Both primary and secondary techniques are designed to remove the petroleum which is readily separated from the mineral grains of the reservoir rocks. However, a considerable quantity of oil is still held by strong capillary forces.

Tertiary extraction techniques have been developed to recover this bound oil. This usually involves injecting petroleum solvents into the reservoir, recovering the resulting solution and finally separating the oil from the solvent. Obviously this method is very expensive and careful planning is essential.

Finally, a more recent advance is that of petroleum mining where the reservoir rock is mined in a similar fashion to any "hard rock" mining operation. The oil is then removed from its host at the Earth's surface by retorting or solvent extraction. This technique is also being considered for mining oil shales.

PETROLEUM ENGINEERING

Once the petroleum deposit has been located, it is then the responsibility of the petroleum engineer to ensure that the maximum amount of petroleum is extracted in the most efficient and economical manner. Not only must the engineer decide on which technique (or combinations of techniques) to use, but also how many wells should be drilled, at what spacing and at what rate must the petroleum be recovered. For example, if the flow of petroleum through the reservoir rock is relatively slow, then one well pumping at a high rate will soon cause flooding of the well by the underlying formation water. The alternative is to accept a slower rate of recovery or drill more production wells each of which extracts the petroleum at the slower rate. The petroleum engineer must be particularly competent in fluid dynamics. Obviously, the ultimate decision is an economic one.

REFERENCES

Giuliano, F. A. (Ed.), 1981. Introduction to Oil and Gas Technology (2nd Ed.). International Human Resources Development Corporation, Boston.

Hobson, G. D. and Tiratsoo, E. N., 1981. Introduction to Petroleum Geology (2nd Ed.). Scientific Press, Beaconsfield.

Hunt, J. M., 1968. How Gas and Oil Form and Migrate. World Oil, October: 140-150.

Hunt, J. M., 1979. Petroleum Geochemistry and Geology. Freeman and Co., San Francisco.

Levorsen, A. I. 1967. Geology of Petroleum (2nd Ed.). Freeman and Co., San Francisco.

Tissot, B. P. and Welte, D. H., 1978. Petroleum Formation and Occurrence. Springer-Verlag, Berlin.

Van Krevelen, D. W., 1961. Coal. Elsevier, New York.

FURTHER READING

TEXTS

American Association of Petroleum Geologists, 1971. Origin of Petroleum. Selected Papers Reprinted from AAPG Bulletin. AAPG Reprint Series No. 1. AAPG, Tulsa.

American Association of Petroleum Geologists, 1974. Origin of Petroleum II. Selected Papers Reprinted from AAPG Bulletin. AAPG Reprint Series No. 9. AAPG, Tulsa.

Tiratsoo, E. N., 1967. Natural Gas. Scientific Press, London.

JOURNALS

American Association Of Petroleum Geologists Bulletin
Australian Petroleum Exploration Association Journal
Oil and Gas Journal
World Oil

Chapter Four

GROUNDWATER

INTRODUCTION

Hydrology is the science that relates to water, specifically with the occurrence of water in the Earth. The hydrologic cycle (Figure 4.1) summarises the basic concepts of hydrology.

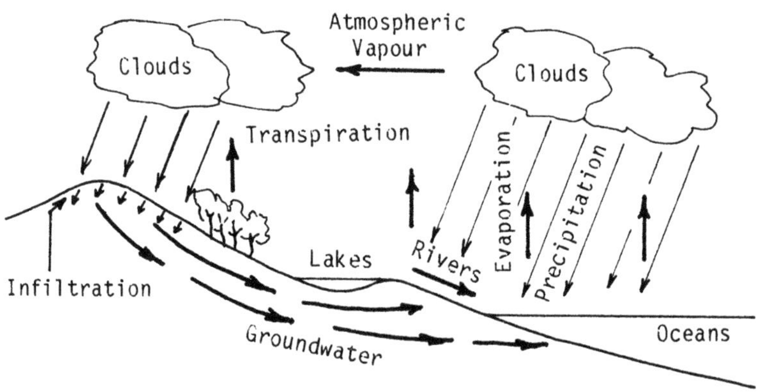

Figure 4.1: A Simplified Hydrologic Cycle.

The water which occurs beneath the Earth's surface is called groundwater and this is the only aspect of hydrology that will be considered further. Concepts fundamental to the understanding of groundwater which are also important in petroleum geology, the migration of ore bearing fluids and engineering geology are:

1. Storage in Earth material
2. Movement through Earth material

3. Geological environments in which groundwater occurs
4. Exploration and development of groundwater resources

There are a number of texts devoted to groundwater including Bouwer (1978), De Wiest (1965) and Domenico (1972). Texts on hydrology in general, such as Ward (1975), also contain chapters on groundwater.

ORIGIN OF GROUNDWATER

Meteoric Water

Meteoric water originates in the atmosphere, falls as precipitation (rain) and becomes groundwater by infiltration.

Connate Water

Connate water is contained in sediments and sedimentary rocks and which was the water in which the sediments were originally deposited. This type of groundwater is subject to controversy because it is difficult to be sure of the origin of water in sediments. This is especially true if the sedimentary rocks are aquifers and large quantities of water flow through them. One obvious example is an artesian basin where, it is generally agreed, the artesian water was meteoric water which fell on to, and infiltrated into, the intake beds. To avoid this problem of determining the origin of this water, the less genetic term "formation water" is gaining wide acceptance.

Juvenile Water

Juvenile water is that water which is considered to have been generated in the interior of the Earth and to have reached the upper levels of the Earth's surface for the first time. It is also called magmatic water. The origin of this type of water is also controversial, especially with the advent of plate tectonic theory and the concept of circulating Earth materials. For example, the water in volcanic eruptions associated with subduction zones may be truly juvenile or have a source in the downgoing subducting plate with its load of marine sediments. It is more likely that the water in volcanic eruptions associated with stable parts of the Earth's crust (such as the Hawaiian Islands) is truly juvenile.

STORAGE

Porosity

Groundwater is stored in the voids or pore spaces of Earth materials. Porosity is the proportion of the volume of the voids or spaces in the material to the total volume of the material. This is expressed as a percentage and is summarised as:

$$\text{Porosity} = \frac{\text{Volume of Voids}}{\text{Total Volume}} \% \quad .....(4.1)$$

It is very difficult to determine a value for porosity as defined above for a number of reasons. One example is where the voids are not connected. Another is where the material is fine grained and molecular forces such as capillary attraction make a large proportion of the water unavailable for use. The normal technique for determining porosity is to weigh a dry sample of the material, soak the material in water for a defined length of time, weigh the "saturated" sample and then calculate the volume of voids by the difference in mass. This technique assumes that all the voids are connected and that upon drying all the water is available for use.

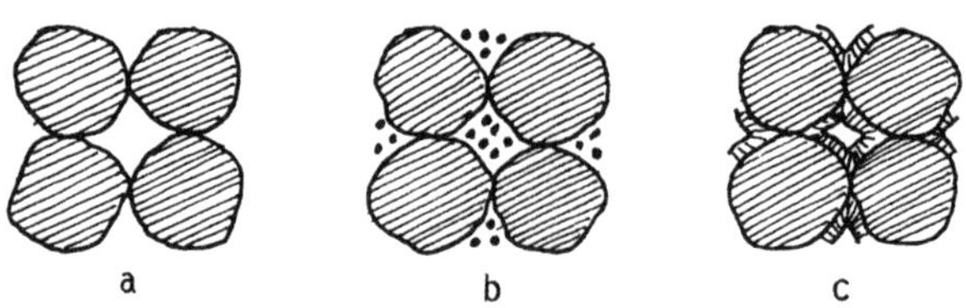

Figure 4.2: Some Types of Porosity.
(a) High Porosity in a Well Sorted Sandstone.
(b) Low Porosity in a Poorly Sorted Sandstone.
(c) Low Porosity in a Well Cemented Sandstone.

Some different ways in which porosity is developed in Earth materials are illustrated in Figure 4.2. Porosity can also be increased by solution and by fractures and/or joints.

Effective Porosity

As a result of the difficulties outlined above, it is more usual to determine the effective porosity. This is a measure of the volume of the voids which are connected and is summarised by:

$$\text{Effective Porosity} = \frac{\text{Volume of Voids Yielding Water}}{\text{Total Volume of Sample}} \quad \text{...(4.2)}$$

The volume of voids that yield water is determined by measuring the volume of water which drains from the sample. Note that this technique does not measure the volume of water held by molecular forces and is therefore of more practical value than total porosity.

Specific Yield

The effective porosity is a measure of another parameter of groundwater called the specific yield. In the laboratory effective porosity is equivalent to specific yield. However, in the field, specific yield is also a measure of the influence of fissures, joints and solution openings.

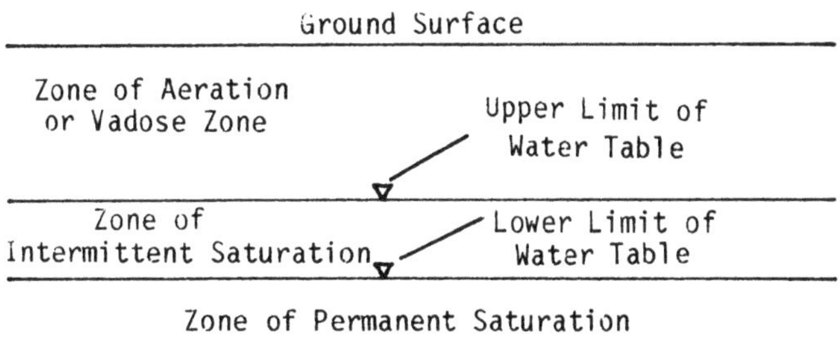

Figure 4.3: Water Table Positions and Zones of Saturation.

The Water Table

The water table is the upper surface of saturation of Earth material by groundwater. This surface changes position depending on the relative amounts of water added to and subtracted from the system. For example, during a rainy season the water table will rise and during drought the water table will fall.

The material above the upper limit of the water table is called the zone of aeration (or unsaturated or vadose zone); the zone between the upper and lower limits of the water table is the zone of intermittent saturation; and the zone beneath the lower limit of the water table is the zone of permanent saturation (Figure 4.3). Such porous and permeable material is called an aquifer.

Water tables can be perched on an aquiclude, or impervious rock (Figure 4.4). When such a water table intersects the land surface it forms a spring. In the example below, the river is above the regional water table and depends on surface runoff from rainfall and the spring flow for its existence. Conversely, if the river bed intersected the regional water table, then the river would be perennial as it would not rely on surface runoff for its existence. Note that the water table tends to rise under hills and fall under valleys.

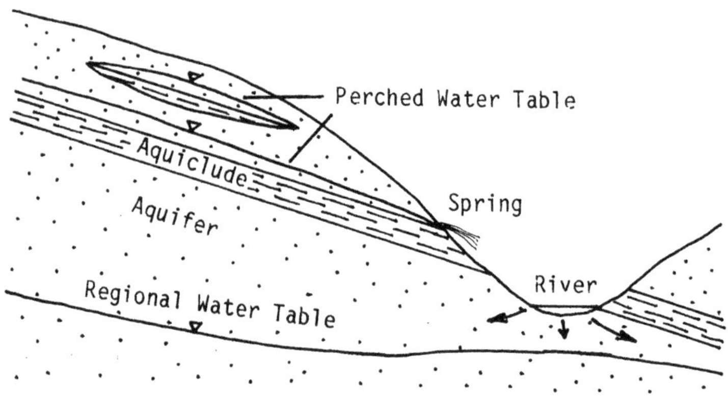

Figure 4.4: Perched Water Tables.

Groundwater Pressure

Groundwater is usually, but not always, under positive pressure. The pressure, or hydraulic, head is the height of the column of water which will balance the groundwater pressure. The elevation head is the height of a point above a given datum. Together, these two give the total head.

The surface which joins the total head in a groundwater regime is called the piezometric surface (after the tubes which are used to measure the pressure head). The hydraulic gradient is the slope of the piezometric surface. In the case of a confined aquifer, the piezometric surface may be above the ground surface and artesian flow will result. If the water rises only part of the distance to the surface it is called sub-artesian water (Figure 4.5).

Note that in Figure 4.5 the wells must be cased for the artesian flow to occur otherwise the water from the lower aquifer will flow into the upper aquifer.

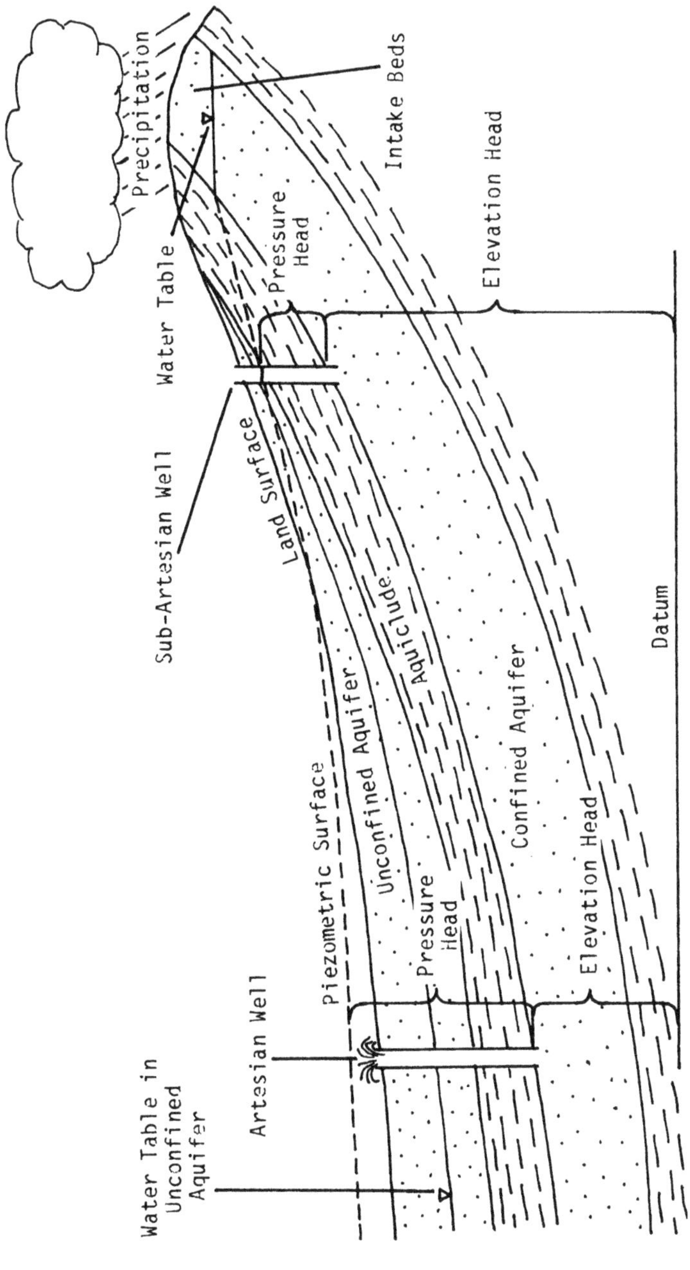

Figure 4.5: Groundwater under Artesian and Sub-Artesian Conditions.

MOVEMENT OF GROUNDWATER

Permeability

Permeability is a measure of the ease with which a fluid will flow through a porous medium. If this material permits relatively rapid movement of the groundwater then it is called an aquifer. Conversely, material which will not allow ready flow is called an aquiclude. The rate of flow of water through very small tubes varies directly as to the hydraulic gradient. If this relationship is applied to water percolating through a porous medium the relationship between the components can be expressed as:

$$v = \frac{Ph}{l} \quad \ldots\ldots\ldots\ldots\ldots\ldots (4.3)$$

where v = the velocity of the water through a column of permeable material
h = the difference in head at the ends of the column
l = the length of the column
P = a constant which depends on the nature of the material

This relationship is often expressed:

$$Q = KAi \quad \ldots\ldots\ldots\ldots\ldots\ldots (4.4)$$

where Q = a measure of the discharge per unit of time
K = the velocity of the water passing through the material
A = the area of the column
i = the hydraulic gradient.

K is therefore a constant (the coefficient of permeability) which varies with different materials and is the permeability in dimensions of velocity. Average values of K for a gravel is 10 cm/sec; a clean sand 0.1 cm/sec and a clayey sand 0.0001 cm/sec. Permeability will vary with temperature because temperature affects the viscosity of the fluid and hence its velocity. If permeability is determined in the laboratory the result will be independent of fissures whereas if the measurements were carried out in the field the result will be related to fissures. This difference gives rise to the concept of perviosity.

Perviosity

Perviosity is related to permeability:

$$\text{Perviosity} = \frac{K(\text{field}) - K(\text{laboratory})}{K(\text{field})} \% \quad ....(4.5)$$

The difference between permeability and perviosity is illustrated by the behaviour of clay. Although all clays are highly porous they are also highly impermeable. However, when clays are dry they are still impermeable but because of shrinkage cracks developed during the drying process they are also pervious. This behaviour is of obvious importance to engineers when the performance of reservoirs is to be predicted for subsequent emptying and filling.

Transmissibility

When investigating an aquifer it is of value to determine the characteristics of the aquifer as a whole. This is done by determining the coefficient of transmissibility. This is a measure of the rate of flow of the groundwater through a vertical strip of aquifer extending the full saturated thickness of the aquifer under a designated hydraulic head.

LOCATING GROUNDWATER

Introduction

A large number of different techniques are available in the search for groundwater. The target is a body of Earth material which is porous, permeable and contains water in quantity and of accepted purity. Some of the techniques are described below.

Topographic Examination

Many of the Earth's surface features give excellent guides to groundwater. Some examples include river valleys (past and present), sand dunes, landslides, fault scarps and colluvial fans.

Vegetation Examination

The water that is available to plants is not necessarily available for pumping, for example, the water contained in clayey soil. However, the type of plant can give some idea of the depth to the water table, especially in arid areas. For example, if grass is present, the water table is within three metres of the surface; for shrubs the depth is within six metres; and for trees the depth of the water table is within 30 metres. If only trees and shrubs are present, it can be assumed that the water table is between three and six metres from the surface. In many arid parts of the world, long lines of very large trees mark water-courses which would otherwise be difficult to identify and which contain water at depth.

Geological Examination

Careful geological examination may locate strata which may provide a perched water table. Deep weathering mantles on tops or bottoms of slopes and the intersections of structures such as faults, folds and dykes may all produce satisfactory aquifers. A knowledge that a particular rock type is a good aquifer combined with an idea of the structure in a folded area will allow predictions of where the aquifer may occur at depth.

Remote Sensing

Apart from using aerial photographs during any of the above techniques, aerial photography may locate areas of evaporation on the ground, these showing up as dark patches on panchromatic film. Satellite imagery can also detect areas of moisture which cause variation in the colour tones of the image. This technique is particularly useful for detecting irrigated marijuana crops in remote arid areas.

ASSESSING GROUNDWATER

Depth to the Water Table

This should be determined by sinking a well, if it is not anticipated to be too deep. More often, the information is obtained by drilling a bore hole. A relatively new technique is to carry out a resistivity survey of the area. The different conductivities of the saturated versus the unsaturated material is usually sufficient for this method to work satisfactorily.

Quality of the Water

The water should be analysed for "normal" soluble salts, other toxic metallic and non-metallic components, solids and bacterial content. If the well density is high, analysis of selected water components may also be of considerable value during mineral exploration.

Productivity of an Aquifer

The productivity, or yield, of a groundwater source is related to a number of factors. When a well is put into production, the extraction of water causes a conical depression of the piezometric surface in the vicinity of the well. This is called drawdown. An interesting phenomenon is that with successive tests the position of the drawdown curve changes, usually becoming less severe (Figure 4.6). This is one method of increasing the yield of a well and is called development.

If the drawdown curve can be limited in severity by development then additional wells can be used without damaging the aquifer. The intersection of the drawdown curve with the normal position of the water table can also determine if the well will suffer any pollution (Figure 4.7).

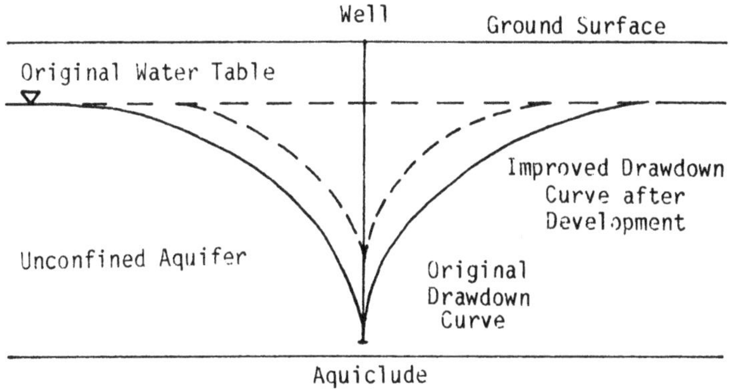

Figure 4.6: The Effect of Development on the Drawdown Curve of a Well.

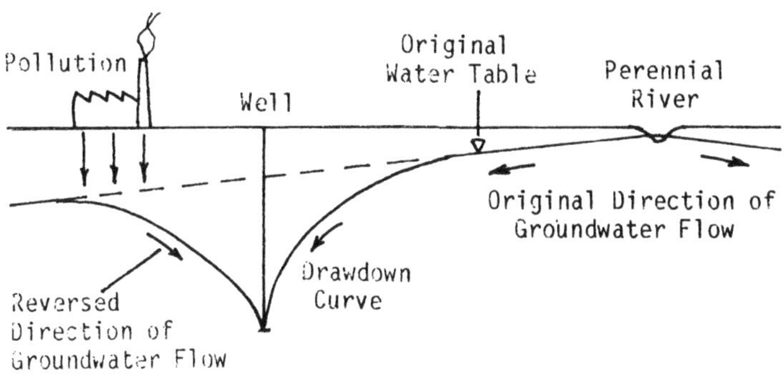

Figure 4.7: The Drawdown Curve of a Well Reversing Groundwater Flow Resulting in Pollution of the Well.

The stability of the well is also important. For example, if the auqifer is an unconsolidated sand it is relatively unstable and will collapse into the well with increasing productivity (pumping). This problem can often be overcome by the use of special screens such as the Johnson well screen which consists of a sub-triangular metal helix supported by vertical rods (Figure 4.8). This type of screen can be easily cleaned.

The ability of the water table to recover to its original level when pumping stops may also be a useful property for assessment.

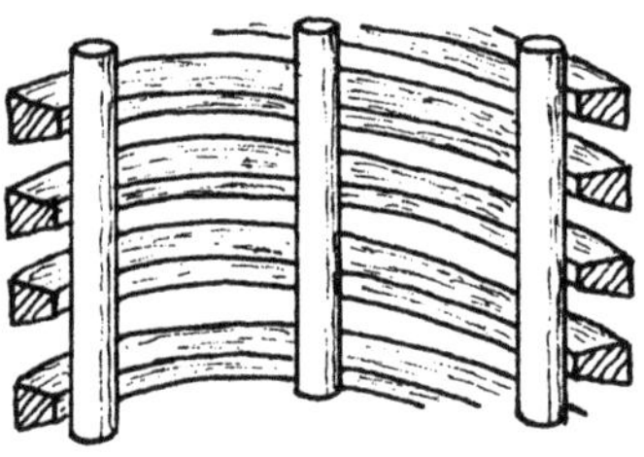

Figure 4.8: The Johnson Well Screen.

WELL DEVELOPMENT

The yield of a well may be markedly improved by well development. As pointed out above, normal pumping of a well usually results in some well development. During pumping the fine particles in the aquifer are drawn into the well and removed. As a result, permeability of the aquifer in the vicinity of the well is improved.

There are a number of artificial methods of well development which include:

1. Surging. This technique involves moving a plunger up and down in the well. The plunger has a one-way flap which results in dragging water out of the aquifer under a greater than normal pressure. As a result, a greater proportion of the fines is removed.

2. Jetting. This technique involves pumping water out of the ends of the rods at a high pressure. This water is directed at right angles to the well sides and also removes

the fines from the aquifer.

Both the above methods result in the removal of the fines from the aquifer in the vicinity of the well and result in a more stable and permeable aquifer.

Other methods include shot firing, which increases the number of fissures; hydraulic fracturing; sand injection (usually in association with the above methods); acidation in carbonate rocks; and placing "dry ice" in the bottom of the well. The ice will evacuate the well instantaneously causing instant drawdown.

REFERENCES

Bouwer, H., 1978. Groundwater Hydrology. McGraw-Hill, New York.

De Wiest, R. J. M., 1965. Geohydrology. John Wiley, New York.

Domenico, P. A., 1972. Concepts and Models in Groundwater Hydrology. McGraw-Hill, New York.

Ward, R. C., (2nd Ed.), 1975. Principles of Hydrology. McGraw-Hill, London.

FURTHER READING

TEXTS

Meinzer, O. E., (Ed.), 1942. Hydrology. McGraw-Hill, New York.

JOURNALS

Journal of Hydrology

Chapter Five

RESOURCE EXPLORATION

INTRODUCTION

The principles described below are generally applicable to any resource, whether it is metallic, non-metallic, petroleum, coal or water, hence the term resource exploration rather than mineral exploration.

The reserves of any deposit are finite, wasting and non-renewable. It is important that there is a continual search for new deposits or extensions to existing deposits so that the continually growing demand for the Earth's resources can be satisfied. The exploration effort must be carefully planned and executed because of the high costs and risks involved in finding a significant deposit. Canadian experience suggests that there is only a 0.2% chance of making a significant find - which may never become a mine. Figure 5.1 shows the relationship between risk and cost during the life of a mine.

This chapter is concerned with the exploration stage of Figure 5.1 - the high risk stage. Two recent texts which treat this topic in detail are Reedman (1979) and Peters (1978). Quick and Buck (1983) discuss planning petroleum exploration programmes and emphasise the importance of statistical and risk methods. The Mining Magazine summarises the exploration and development of a mine in most issues. Before the preliminary reconnaissance stage can be commenced, the whole exploration programme must be planned.

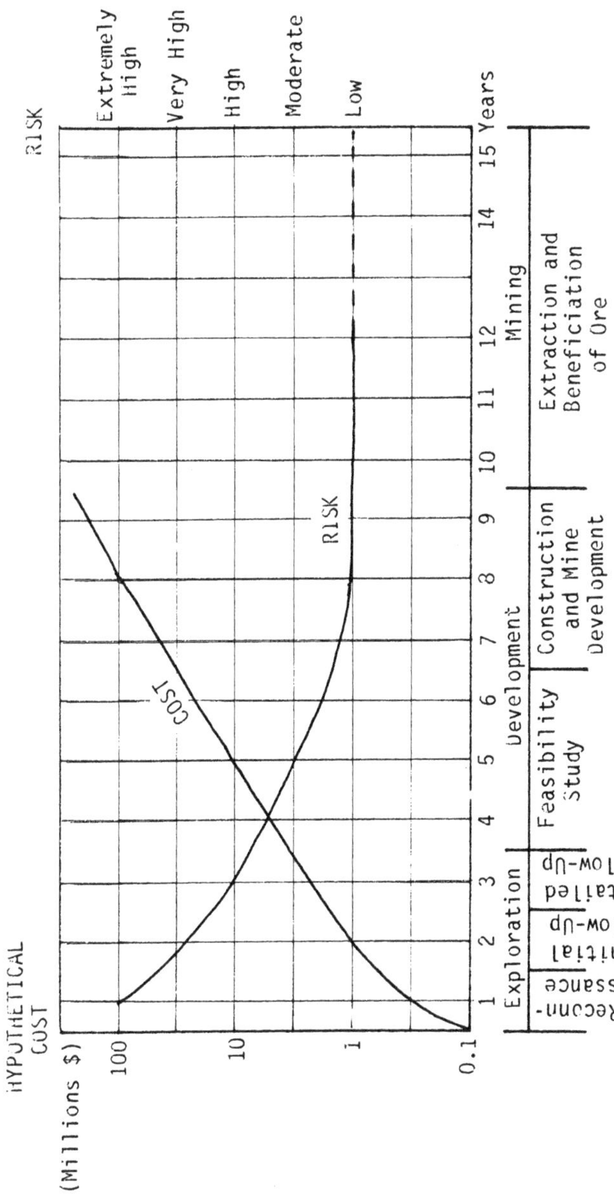

Figure 5.1: The Relationships Between Time, Cost, Risk and Stages of Exploration, Development and Mining. (Data from Reedman, 1979).

PLANNING, OR THE DEVELOPMENT OF AN EXPLORATION PHILOSOPHY

Economic and Political Stability of the Country

These factors must be carefully considered and analysed before any substantial sums of money are spent. Particular care must be taken in assessing the attitudes of governments and trade unions to private enterprise. Points to consider include:

1. Degree of local equity desired
2. Taxation and royalty rates
3. Mining law
4. Controls on the import of capital and the export of profits

High taxes and royalties, although not desirable, are of less concern to mining companies than rapid changes to these factors. Stability is required so that long term planning can be brought to fruition. High costs and taxes are acceptable provided that the company can make a reasonable return on the risk money of exploration and the capital invested in the mine and infrastructure. However, if the ground rules change during the life of the mine, the company can stand to lose not only profits but capital. The worst prospect that a company can face is that of nationalisation of its assets with little, or no, compensation.

Which Minerals?

As a generalisation, the more vertically integrated the company the easier it is to answer this question. Companies in the aluminium and petroleum industries are examples. These companies are typically involved in the exploration, mining, refining, and marketing of only one resource. In contrast, horizontally integrated companies, such as those interested in only one or two aspects of resource exploitation (for example, exploration and mining) may consider any commodity on its merits as the opportunity presents itself.

Of course, some commodities become popular exploration targets because of economic and political events. The nickel boom of the late 1960s was fuelled to a large extent by long continuing labour disputes in the major nickel mines in Canada. The 1970s saw a dramatic increase in exploration for fossil fuels because of pclitical events in the Middle East. The early 1980s "gold rush" in Australia was principally due to the weaker Australian dollar compared to the US dollar. This resulted in a high price for gold as expressed in Australian dollars and potentially large profits for gold producers.

Optimum Size of Deposits

This is a very important consideration not only with respect to the company's ability to fund the development of a mine but also with respect to the design of the exploration programme. Small companies tend to explore for small high grade deposits while large companies explore for larger and lower grade deposits.

If, by chance, a company finds a large deposit, it may not have the financial ability to develop the deposit and may have to sell off the deposit. An alternative is to invite participation from a larger company to provide financial or technical assistance. In return, the invited company acquires an equity in the deposit. The smaller company has "farmed out" equity to the larger company which has "farmed in".

Ideally, the company decides the amount of capital that can be made available for mine development. The amount of profit required after all costs, including taxes, royalties and capital repayment are recovered, is then determined. A certain style of mineralisation and ore grade is then assumed and the size of the ideal deposit calculated.

Reconnaissance exploration should be designed so that deposits smaller than the ideal size are not detected whilst larger deposits are located. Thus money is not wasted on detecting and following up deposits which would not be economic for the company to develop.

Current Concepts of Ore Genesis

The importance of the exploration team members being made aware of the current concepts of ore genesis cannot be emphasised enough. Theories of the formation of mineral deposits change as more geological information becomes available and a knowledge of these changes directs the aware geologist to areas of potential mineralisation.

One clear example of the influence a change in theory may have is that a few decades ago it was almost mandatory for the presence of plutonic igneous rocks for base metal mineralisation; now sedimentary rocks containing organic carbon with or without volcanic and evaporite rocks is a favoured association.

Literature Search
A careful, detailed review of the literature is fundamental to selection of favourable areas for exploration. Although literature searches may be started in the company's library and move out into Geological Survey, University and other libraries, the tendency today is to use one of the many computer literature data bases. The customer decides on the parameters of the search such as locality, mineral type, exploration methods, and within a short time, usually 24 hours, the investigator has the basis of a very detailed bibliography within those limits. There are other similar systems both in Australia and overseas.

Selection of Areas of
Favourable Geological Environment
This is the end result of the previously described steps, particularly the literature search and awareness of current ore genesis theory. For example, the search for diamonds in north-west Australia was essentially based on present-day understanding of kimberlite formation and the concept of continental drift. That is, if that part of Australia is geologically similar to diamondiferous Africa, then logically, kimberlite pipes could be expected to be found in north-west Australia and that some of these may be diamondiferous.

SCANNING OR RECONNAISSANCE EXPLORATION

Title
It is of paramount importance that the company holds the appropriate mineral titles. An obvious point, but records of Mining Warden's Courts contain too many cases of disputed titles. A carefully kept file of titles and their renewal dates under competent and diligent supervision should ensure that no problems will arise. The correct marking and description of areas is also essential. Prospective areas are highly sought after and there is often fierce competition to acquire titles. Figure 5.2 illustrates how little available ground exists for pegging during a resource boom.

It follows that a thorough understanding of the relevant mining law is important. In Australia, and many other countries with a federal political system, this can be a time consuming problem as each state has different mining laws. The Federal

Government can also become involved with mineral titles directly. For example, it is usually the responsibility of the Federal Government to pass laws governing coastal waters. In some countries mineral rights are held by landowners. Negotiations must then be undertaken before beginning exploration.

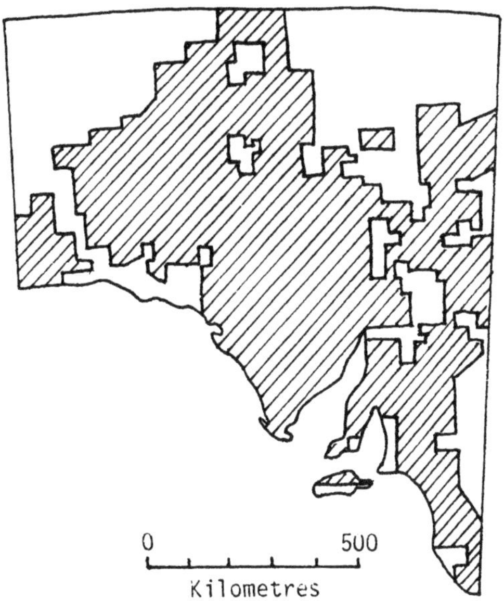

Figure 5.2: The Area of South Australia Covered by Mineral Exploration Licences in June, 1981.

Reconnaissance exploration normally requires relatively large areas to be held under such titles as "Exploration Licence" or "Temporary Reserve" or something similar. All of these are expensive to hold so one of the main objectives of reconnaissance exploration is to select the best target areas by eliminating the least attractive ground as soon as possible. Some petroleum exploration licences actually insist on the surrender of a portion of the licence area on a yearly basis.

Aerial Photographic Interpretation

The majority of countries are now covered by aerial photographic surveys and it is relatively inexpensive to buy photographs of the area under investigation. Examination of the photographs stereoscopically should be carried out before any field work. Base maps may have to be prepared especially if the already published maps have an unsatisfactory scale or if the area is remote and no base maps exist. Photogeological interpretation should be made with the assistance of satellite imagery if possible.

Geological Mapping

No matter how good the content of the literature, how good the previous work and how good the aerial photographic interpretation, there is no substitute for field mapping. Particular attention should be given to old mine workings and attempts should be made to determine the source of any geophysical anomalies recorded by previous work. Field parties should be small and highly mobile.

Prospecting

Persons familiar with the resource and/or the area selected and who do not have academic qualifications but a wealth of practical experience are often valuable during reconnaissance exploration. If given a specific task, such as collecting specimens from all "ironstone" outcrops, a great amount of information can be obtained at relatively little cost. Most mines were found by prospectors and even during the nickel boom in Western Australia in the late 1960s prospectors proved most valuable in locating mineralisation.

Aerial Geophysical Surveys

If previously flown aeromagnetic surveys are not considered to be satisfactory because of height flown, line spacing or less sophisticated technology, it is usually justifiable to have covered the area by a well planned aeromagnetic survey. Regional surveys are primarily used to determine the regional geology. Crystalline complexes can be distinguished from sedimentary basins, basement structures can be determined and highly magnetic rocks, such as banded iron formations (BIFs), can be used to assist in interpreting the structure. Airborne scintillometer surveys may also provide useful information at this stage.

Regional Geochemical Surveys

These are usually stream sediment surveys with a low density of sampling so that a regional geochemical relief map can be prepared. In areas of little or no drainage, regional geochemical soil surveys on a grid system are often used. This latter method has proved useful in locating diamond pipes in

Africa and nickel prospects in Australia.

Assessment

The above work usually results in the definition of one or more anomalies. At the conclusion of the field work (and preferably as an ongoing review) decisions must be taken as to which areas (if any) offer potential for containing an economic deposit. Obviously, economics plays almost as important a role as geology at this stage but it must be recognised that although the risk decreases at the next stage, the cost doubles. The most common approach is to list the favourable areas in decreasing order of estimated importance and to spend a fixed amount of money on each.

INITIAL FOLLOW-UP

Detailed Geological Mapping

The favourable areas defined by the reconnaissance stage are now mapped at a larger scale, often by using enlarged aerial photographs as base maps. The scale should be large enough so that as much geological information as possible may be plotted. In most exploration programmes, this mapping is carried out in conjunction with detailed geophysics and geochemistry. Field camps can be more permanent than at the reconnaissance stage.

Detailed Geophysics

The importance of deciding the minimum target size during the development of philosophy and planning stage now becomes obvious. Line spacing and sampling interval should be planned so that at least one line will cross the expected minimum anomaly at right angles and also that at least one sample station will detect the anomaly if it exists.

Both ground and low level airborne (usually helicopter) geophysical surveys may be used, the techniques depending on the target and the geology. For example, in base metal exploration, ground magnetics, electromagnetic, scintillometer, electrical and gravity techniques are common whilst in coal and petroleum exploration magnetic, gravity and seismic techniques are used.

Detailed Geochemistry

Grid spacings would generally be similar to those used for the detailed geophysical surveys. Soil and vegetation sampling are the methods commonly used at this stage, although in favourable areas such as glaciated terrains rock sampling may be more useful.

Limited Drilling

Some shallow drilling may be carried out. This drilling is designed to determine the source of geophysical and geochemical anomalies or to obtain geochemical samples from below weathered or transported overburden. As the initial follow-up programme progresses drilling targets specific for mineralisation will hopefully be delineated. All targets which conform to the minimum size requirements should be drilled. Petroleum exploration may require the drilling of stratigraphic wells.

Assessment

Large areas should now be discarded. The detailed geological mapping, geophysics, geochemistry and drilling will now allow selection of areas with potential for ore grade mineralisation. These areas are normally pegged as mineral claims at this stage.

DETAILED FOLLOW-UP

Drilling

If the programme has proceeded to this stage then detailed diamond drilling (or appraisal wells, in the case of petroleum exploration) should be carried out to determine the size and grade of the deposit.

Limited Metallurgical Testing

Samples of drill core, or more effectively, samples from a trial shaft, are investigated in laboratories to determine suitability of the ore to the processes of mining, comminution, separation, concentration and smelting.

Assessment

The end of the exploration programme and a major decision watershed has been reached. If the deposit is promising then a feasibility study is commenced. If not, the area is relinquished.

OTHER TARGETS

The above discussion is of "grass roots" exploration. Exploration can also be directed to:

1. Extending known deposits. Once an ore body has been found and extraction begins, the immediately surrounding area must be explored for continuation or repetition of the mineralisation. Exploration usually concentrates on structural methods to determine if the mineralisation occurs along strike, down dip, as en-echelon bodies or as off-set faulted bodies
2. Reviving known deposits. Some deposits may be small and/or abandoned because of economic considerations. Such deposits may be re-opened because of changes in technology and/or the price of the commodity. For example, many small abandoned gold mines in Australia were re-opened in the early 1980s because the gold price was relatively high; bulk, low-grade mining operations are used instead of costly, selective mining of small tonnages of high grade ore; and the technological development of the relatively inexpensive carbon-in-pulp technique for gold recovery

REFERENCES

Peters, W. C., 1978. Exploration and Mining Geology. John Wiley, New York.

Quick, A. N. and Buck, N. A., 1983. Strategic Planning for Exploration Management. D. Reidel Publishing Co., Dordrecht.

Reedman, J. H., 1979. Techniques in Mineral Exploration. Applied Science, London.

FURTHER READING

TEXTS

Lawrence, L. J. (Ed.), 1970. Exploration and Mining Geology. 8th.Common. Mining and Metall. Congress, Aust. and N.Z., Aus.I.M.M., Melbourne.

JOURNALS

Australian Mining
Mining Journal
Mining Magazine
Mining Annual Review

Chapter Six

EXPLORATION GEOPHYSICS

INTRODUCTION

Geophysical exploration is the method of searching for concealed deposits of useful minerals, petroleum or water by taking physical measurements at the Earth's surface. The information on the physical properties determined, if properly interpreted in terms of geology, can be used to locate the concealed deposits.

Such prospecting is the applied branch of geophysics and the properties which are most useful are:

1. Density
2. Magnetism
3. Elasticity
4. Electrical
5. Radioactivity

and the corresponding branches of applied geophysics are:

1. Gravity
2. Magnetic
3. Seismic
4. Electrical
5. Radiometric

Classic texts on applied geophysics include Dobrin (1976) and Parasnis (1979). Griffiths and King (1981) is a good introductory text. Hansen and others (1966) present a large selection of geophysical exploration case histories.

GRAVITY

Principles

In many respects, gravity surveys are simple to interpret because a natural field is measured and simple fundamental principles are used in the interpretation. Newton's Law of Universal Gravitation expresses the mutual attractive force (F) between two masses, m_1 and m_2 separated by a distance r as:

$$F = G\frac{m_1 m_2}{r^2} \qquad (6.1)$$

where G is the Universal Gravitational Constant which has a value of 6.670 x 10^{-11} $Nm^2\,kg^{-2}$.

Also, from Newton's Second Law of Motion:

$$F = ma \qquad (6.2)$$

In the case of a small body falling (that is, being accelerated by the Earth's gravitational field, g) Equation 6.2 may be written as:

$$F = mg \qquad (6.3)$$

and

$$F = G\frac{m_1 m_2}{r^2} \qquad (6.4)$$

where m^1 is the mass of the Earth, m^2 is the mass of the small body and r is the radius of the Earth.

Combining Equations 6.3 and 6.4:

$$g = G\frac{m_1}{r^2} \qquad (6.5)$$

g is the force of gravity and is expressed in units of acceleration called gravity units ($\mu m\ s^{-2}$). The older unit is a gal (1cm sec^{-2}) but it is more convenient to use the unit milligal (one thousandth of a gal) because gravity anomalies are very small. One milligal is thus equal to 10 gravity units. Modern gravity meters (or gravimeters) can measure to an accuracy of 0.1 gravity units. The gravitational acceleration at the Earth's surface is about 9.8 m sec-2 (that is, about 9800 000 gravity units).

When a gravity survey is carried out the value of g depends on a number of factors apart from variations in density. It is necessary to make several corrections to the recorded values of g as measured at each station.

Corrections

Corrections for Latitude. This correction is necessary because the Earth is flattened at the poles and bulges at the equator. The correction is made up of three components:

1. Variation in radius. The radius of the Earth is greater at the equator than at the poles. Therefore, this component tends to increase g with increasing latitude.
2. The centrifugal acceleration of rotation gc. gc acts in opposition to g so that the effect is to decrease the value of g with decreasing latitude.
3. Variations in mass. The greater radius at the equator means that there is more mass at the equator and therefore the effect is to increase g with decreasing latitude.

The first two components reduce the value of g at the equator in relation to the poles by approximately 100,000 gravity units but the last component reduces this effect to about 53,000 gravity units. When gravity surveys cover large areas the nett correction for latitude can be determined from prepared tables. Smaller and more detailed surveys can be corrected by using the International Gravity Formula:

$$g_\lambda = 9.78049\,(1 + 0.0052884 \sin^2\lambda - 0.0000058 \sin^2\lambda)\ \mathrm{ms}^{-2} \quad \ldots.(6.7)$$

where λ is the latitude. At 45° latitude the correction is about one gravity unit per 12 metres of distance in a north-south direction.

Corrections for Elevation. Two corrections are needed:

1. The Free Air Effect. This is simply due to variations in vertical height and therefore variations in r in Equation 6.5. The effect is -3.086 gravity units per metre. That is, with an increase in height the correction is positive.
2. The Bouguer Effect. This is due to the increased mass associated with the increase in elevation. g increases by 0.4185ρ grav-

ity units per metre increase in elevation where p is the density of the material. The correction is therefore negative.

Note that the magnitude of the corrections dictates that accurate surveying of the height of stations is necessary (to within 0.03 metre). This is also the reason why the results of airborne gravity surveys are not usually reliable given the unstable platform that the aircraft provides. Recent advances in computer applications suggest however that regional airborne gravity surveys may be feasible.

Correction for Topography. This is applied where there is considerable variation in relief. A mountain, for example, will affect g depending on its distance and elevation from the station. The correction is very complicated and time consuming.

The Bouguer Anomaly. After all the above corrections have been made to the observed value of g, the result is Bouguer anomaly (Δg) which is simply the result of variations in subsurface density. Figure 6.1 illustrates the expected Bouguer anomalies with different rock densities.

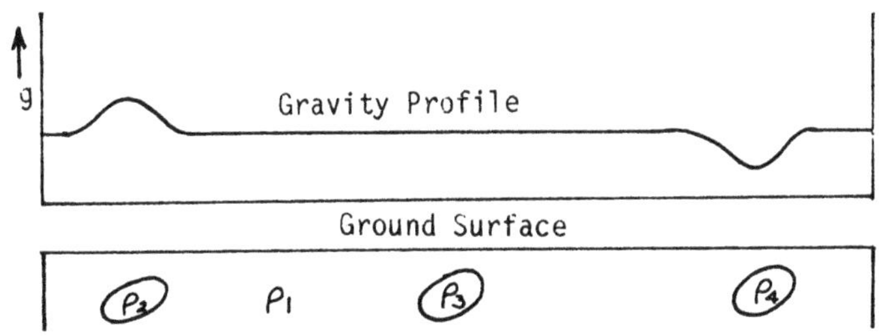

Figure 6.1: Bouguer Anomalies and Density Contrasts.
(a) ρ_2 greater than ρ_1, (b) ρ_3 equal to ρ_1, (c) ρ_4 less than ρ_1.

Residual gravity maps are of great importance when searching for masses and/or structures above the basement. Many Bouguer anomaly maps show a distinct regional gradient in values across the map. These gradients are usually due to an inclined basement surface. If this regional gradient is subtracted from the Bougueranomaly a residual gravity map is produced. Figure 6.2 illustrates how this technique can highlight otherwise very subtle anomalies.

Interpretation

Shape. The pattern of the contoured values of g are interpreted as simple geometric shapes. Concentric contour lines are considered to represent a point source or a sphere and elongate anomalies can be considered to represent a slab or horizontal cylinder.

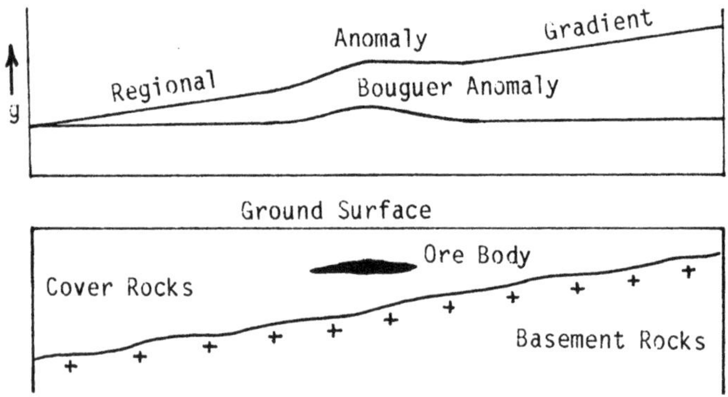

Figure 6.2: Bouguer Anomaly and Residual Anomaly Profiles.

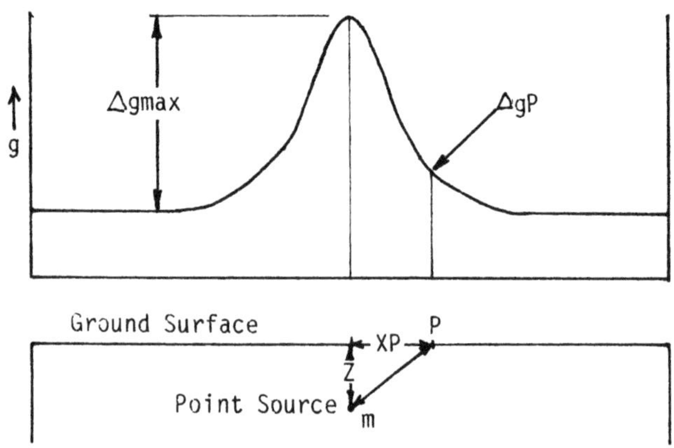

Figure 6.3: The Relationship of the Shape of a Gravity Anomaly to the Depth of a Point Source.

Depth. A simple case of a three dimensional source of a gravity anomaly is a spherical massive sulfide body with its centre as a point of mass. The relationship of the gravity profile over such a point source is illustrated in Figure 6.3.

The vertical component of Δg (the Bouguer value) at a point P which is at a distance XP from the point of the maximum anomaly (vertically over the source m) is:

$$\Delta g_p = G\frac{m}{z^2} \cdot \frac{1}{\left(\frac{1 + xp^2}{z^2}\right)^{3/2}} \quad \text{........(6.8)}$$

and the relationship between Δg and Δg max is:

$$\frac{\Delta g_p}{\Delta g_{max}} = \left(1 + \frac{(x_p)^2}{z^2}\right)^{-3/2} \quad \text{........(6.9)}$$

From these equations it is therefore possible to calculate the depth (Z) of the source of the anomaly (m). There are two simple rules which enable rapid calculations of very close approximations to true depth. The first assumes that XP = Z when:

$$\frac{\Delta g_p}{\Delta g_{max}} = 2^{-3/2} \quad \text{..................(6.10)}$$

that is, when the anomaly has fallen to one third of the maximum anomaly then X = Z. The second assumes that XP = 0.767Z when:

$$\frac{\Delta g_p}{\Delta g_{max}} = \frac{1}{2} \quad \text{....................(6.11)}$$

that is, when the anomaly has fallen to one half of its maximum value then X = 0.767Z.

Interpretation of two dimensional sources of gravity anomalies is considerably more complicated.

Mass. Once the depth to the source of the anomaly (in the case of point sources, Z) is known and as Δg_{max} is known, then:

$$\Delta g_{max} = \frac{G.m}{z^2} \quad \text{....................(6.12)}$$

from which the excess mass (m) of the body can be calculated. Remember that a gravity anomaly is due to density and mass differences.

If the true mass of the source of the gravity anomaly is M and the mass difference is m, then their relationship is:

$$M = m.\ \frac{\rho_2}{\rho_2 - \rho_1} \quad \text{..........(6.13)}$$

where ρ_1 and ρ_2 are the densities of the surrounding rock mass and the source of the anomaly respectively.

These calculations can be carried out in reverse before a survey to predict the size of the anomaly which would be associated with a body of the desired size.

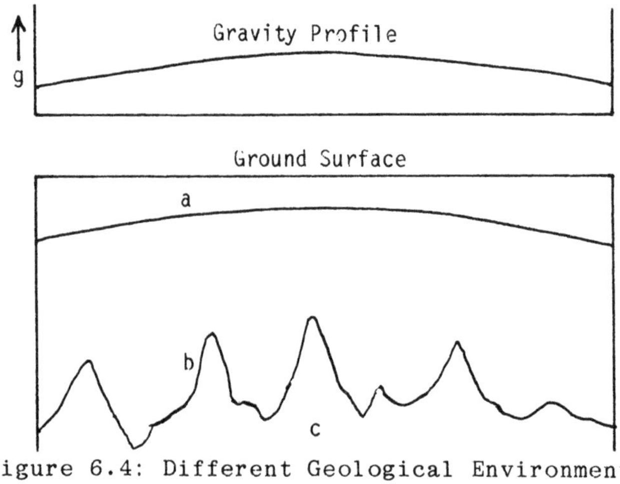

Figure 6.4: Different Geological Environments Producing an Identical Gravity Profile.
(a) Shallow Regular Basement Topography.
(b) Deep Irregular Basement Topography.
(c) Deep Point Source.

Size. The above discussion assumed a point source of the anomaly. Surveys are carried out to locate three dimensional bodies and those with a point source are spheres. The relevant equation is:

$$m = \frac{4}{3}\pi r^3(\rho_2 - \rho_1) \quad \text{......(6.14)}$$

where m = mass excess
r = radius of the sphere

ρ_1 = density of the mass
ρ_2 = density of the surrounding material

If the previous calculations have been made, then m, ρ_1 and ρ_2 are known and it is a simple matter to calculate r and hence the size of the body.

Modelling. An alternative interpretive method is modelling. Various models of the subsurface geology are assumed and their predicted gravity effects are compared with the actual gravity measurements. The model of best fit is considered to be the most probable one. However, a number of quite different geological situations can produce identical gravity profiles (Figure 6.4).

MAGNETIC

Principles
Magnetic surveys are the easiest to carry out and are used to prospect for minerals (directly) and petroleum (indirectly). The basis of this method is that rocks are capable of influencing (distorting) the Earth's magnetic field. By measuring these distortions and interpreting the results in the light of geological knowledge considerable additional information about the subsurface geology can be obtained. Magnetic surveys, however, are used mainly in regional reconnaissance surveys as a geological mapping tool. A rather unique application of magnetic surveys is that of the world's navies. Submarines have detailed magnetic maps of the oceans so that they can hide in magnetically "noisy" portions of the ocean floor. This makes their detection more difficult by sophisticated magnetometers in anti-submarine aircraft.

The following concepts will help in understanding the principles of interpretation:

1. Magnetic Poles. Magnetic material has both north seeking (positive) and south seeking (negative) poles.
2. Magnetic Force. The force between two poles varies inversely as the square of their distance apart, or:

$$F = \frac{1}{\mu} \frac{m_1 m_2}{d^2} \quad \text{................(6.15)}$$

where μ is a constant depending on the

medium in which the poles exist (in air $\mu = 1$). Note the similarity between Equation 6.15 and Equation 6.1 (for gravity)

3. Magnetic Field. The magnetic field is the magnetic influence around a magnetic particle as illustrated in Figure 6.5. In the SI system magnetic fields are measured in nanoteslas where one nanotesla equals one gamma or 10^{-5} gauss in the e.m.u. system of measurement.

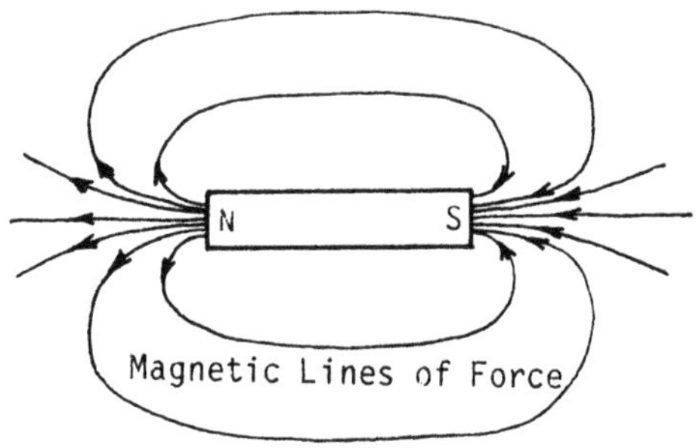

Figure 6.5: The Magnetic Field Around a Bar Magnet

4. Magnetic Induction. This phenomenon occurs when magnetic poles are induced in a material by an external magnetic field. The intensity of magnetism may be considered to be the induced pole strength per unit area along a surface normal to the inducing field. It is the intensity of magnetism which is normally measured by magnetic surveys.
5. Magnetic Susceptibility. This describes the ease with which induced magnetism can be caused in a material by an external magnetic field. If the particles in the material line up with their long dimensions parallel to the direction of the external field then the material is said to be paramagnetic. If the particles line up with their long dimensions across the field the material is said to be diamagnetic. Materials and minerals that acquire magnetic properties of any size are all paramagnetic, for example, magnetite, ilmenite and pyrrhotite.
6. Residual Magnetism. This is often called

fossil, remanent or palaeomagnetism. The magnetism was aquired during the formation of the rocks when magnetic particles aligned themselves with the Earth's magnetic field at that time. In the case of igneous rocks the residual magnetism was acquired when the rocks cooled below the Curie point (the temperature above which material loses its permanent magnetism). Sedimentary rocks, particularly sandstones, often show residual magnetism where water deposition allowed individual magnetic minerals to line up with the Earth's magnetic field at the time of deposition. Residual magnetism introduces a complication to the interpretation of magnetic survey results which does not apply to gravity surveys. In fact magnetic surveys measure the resultant of induced and residual magnetism. For a detailed discussion of the origins of residual magnetism see any standard geophysical text.

The Earth's Magnetic Field

The Earth's magnetic field may be explained by an imaginary magnet at the centre of the Earth with an orientation of magnetic south pole towards the Earth's north geographical pole (Figure 6.6). The

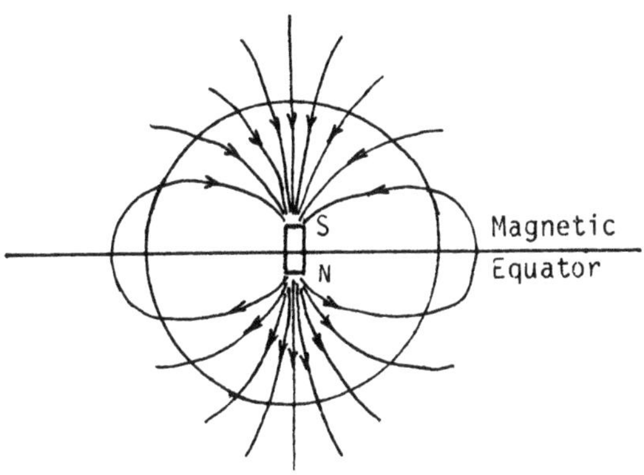

Figure 6.6: The Earth's Magnetic Field.

distribution of the magnetic lines of force results in higher magnetic strength at the poles (63,000 nanoteslas) compared to that at the equator (31,000 nanoteslas). The angle that the field makes with the horizontal at any point on the Earth's surface is called the dip or inclination.

The Earth's magnetic field varies in three ways:

1. Secular variation. A slow and continuous variation and for the purposes of magnetic surveys can be ignored.
2. Diurnal variation. This is a change in the Earth's field strength of about 25 nanoteslas during the day. It is probably due to the relative position of the point of measurement and the Sun (that is, it is a function of the rotation of the Earth). If a great accuracy is required for a magnetic survey then the diurnal variation must be measured constantly and the survey results modified accordingly.
3. Magnetic storms. There are periods when the Earth's magnetic field may vary by over 1000 nanoteslas over hours or days. This effect is probably due to sunspot activity. Magnetic surveys are very difficult if not impossible to carry out during such storms.

Interpretation of Magnetic Anomalies

Conceptually, the source of a magnetic anomaly is a large magnet buried in the Earth whose field affects the Earth's magnetic field. It is this effect which is measured during a magnetic survey.

The magnitude of a magnetic anomaly depends on:

1. The size, shape and attitude of the body.
2. The direction and intensity of the Earth's magnetic field at that point.
3. The magnetic susceptibility contrast between the body and its surroundings.
4. The remanent or palaeomagnetism which is generally unknown and thus makes interpretation complicated.

Magnetic anomalies are two or three dimensional as are the gravity anomalies. Note that for convenience in the southern hemisphere the anomalies are plotted upwards, that is, as if they were positive.

The method of interpreting a magnetic anomaly is identical to that of interpreting a gravity anomaly provided that only one pole is considered.

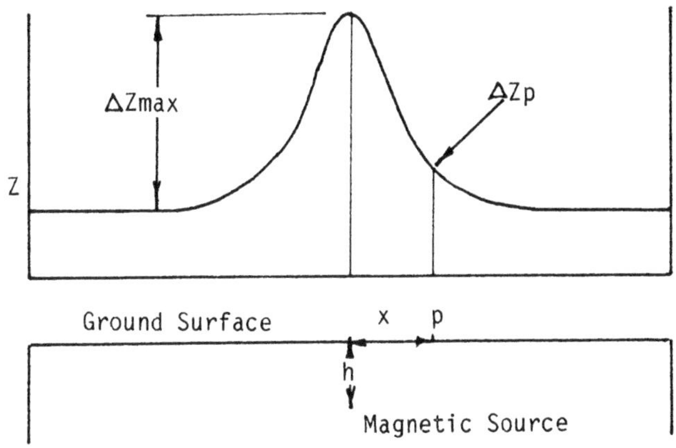

Figure 6.7: The Shape of a Magnetic Anomaly Over a Point Source of Magnetism.

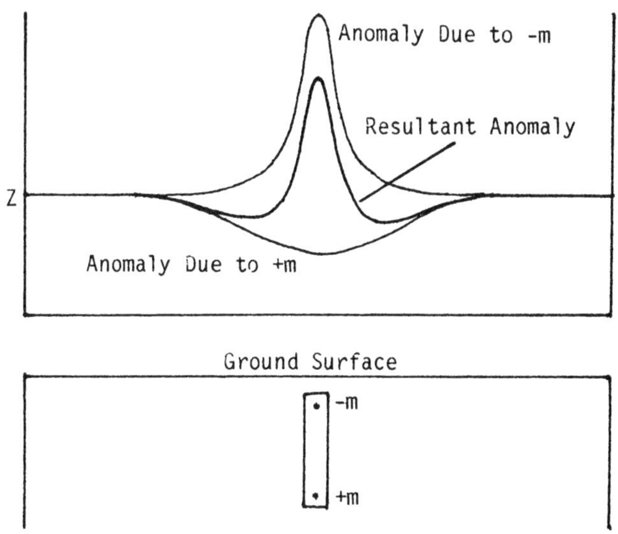

Figure 6.8: The Anomaly Resulting from a Vertical Magnetic Body of Finite Size.

That is, the opposite pole is at such a depth that

its effect is negligible and also that the remanent magnetism is negligible (Figure 6.7 and Equation 6.16). Note that the value of pole strength is

$$\frac{\Delta Z_p}{\Delta Z_{max}} = \left(1 + \frac{x^2}{h^2}\right)^{-3/2} \qquad (6.16)$$

not needed for interpretation. If both poles of the body affect the anomaly then negative "wings" are developed. Figure 6.8. shows such effects for a vertical body and Figure 6.9 for an inclined body. Interpretation is more complicated when the effects of the variation of magnetic dip with latitude are considered.

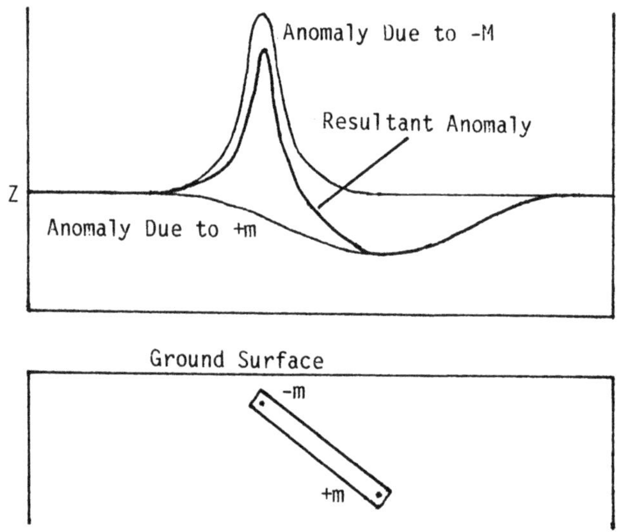

Figure 6.9: The Anomaly Resulting from an Inclined Magnetic Body of Finite Size.

SEISMIC

Principles

Seismic prospecting is the most widely used of all the geophysical methods. It is mainly used in petroleum exploration and to a lesser extent in foundation engineering investigations.

In general terms the seismic method involves inputting energy into the Earth, usually by setting off an explosion. The resulting shock waves are detected and their travel time determined. Depending on whether the waves have been refracted or reflected different interpretive methods are used to determine the subsurface geology.

When a charge is fired at or near the surface of the Earth three types of waves result:

1. Surface waves which absorb most of the energy
2. Reflected waves
3. Refracted waves

Reflected and refracted waves form at a subsurface boundary between materials of different elasticity and hence different ease with which the waves are transmitted. For example Figure 6.10 illustrates Snell's Law which is:

$$\frac{\sin\theta 1}{V1} = \frac{\sin\theta 2}{V2} \quad \ldots\ldots\ldots\ldots\ldots\ldots (6.17)$$

where V is the velocity at which the waves are transmitted through the different rock layers.

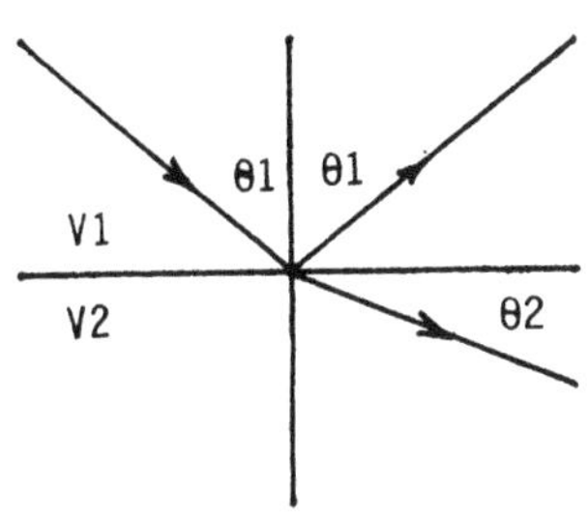

Figure 6.10: Reflected and Refracted Waves in a Two Layered Environment.

If the incident waves aproach the boundary at what is called the critical angle (θc) then the wave travels along the interface as shown in Figure 6.11 and Sinθc is V1/V2. However, as the wave is travelling along the boundary it releases energy as shown in Figure 6.11 and these waves can be detected at the surface.

The shock waves are composed of primary, secondary and Raleigh waves. To avoid confusion during interpretation only the first arrival waves (the primary waves) are considered. The velocity of the waves in rocks is not determined by the rocks elasticity alone and for this reason different layers are identified by their velocities rather than lithology.

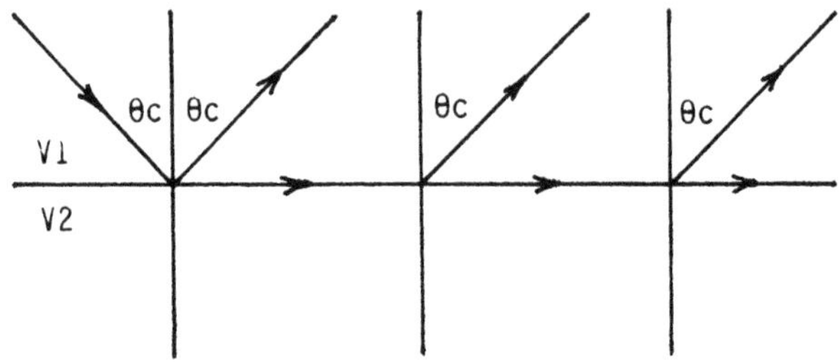

Figure 6.11: The Path of the Wave at the Critical Angle.

<u>The Seismic Refraction Method</u>
The refraction method was the first to be used in seismic prospecting although the reflection method is now more widely used. Refraction surveys have the advantage over reflection surveys in that seismic velocities can be determined directly for the various layers in the crust.

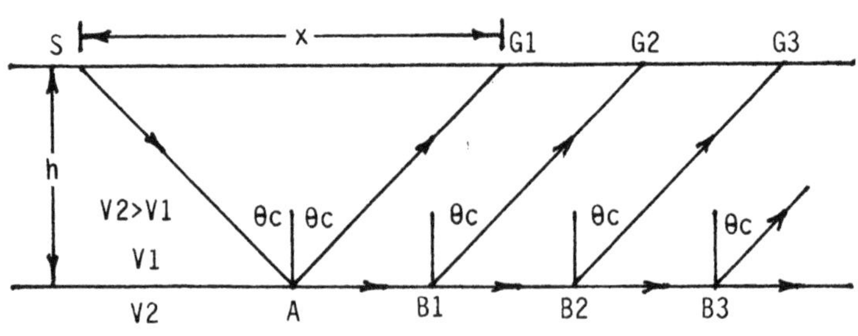

Figure 6.12: The Principles of a Seismic Refraction Survey.

Figure 6.12 illustrates a simple two layer refraction survey. S is the shock point, G1, G2, on so on are geophone positions and h is the depth to the boundary. In a general case, a geophone, G, will detect the arrival of both the direct wave SG and the refracted wave SABG. If x is small the direct wave will be recorded first. However as x increases a point will be reached where the refracted wave, which has travelled in the upper surface of the higher velocity material, will arrive before the direct wave.

It can be shown that:

Direct arrival time $T_1 = \frac{x}{V_1}$ ....(6.18)

and Refracted arrival time:

$$T_2 = \frac{x}{V_2} + 2h\sqrt{\frac{1}{V_1^2} - \frac{1}{V_1^2}} \quad \text{.....(6.19)}$$

Arrival times are plotted against geophone distances from the shock point in Figure 6.13. It

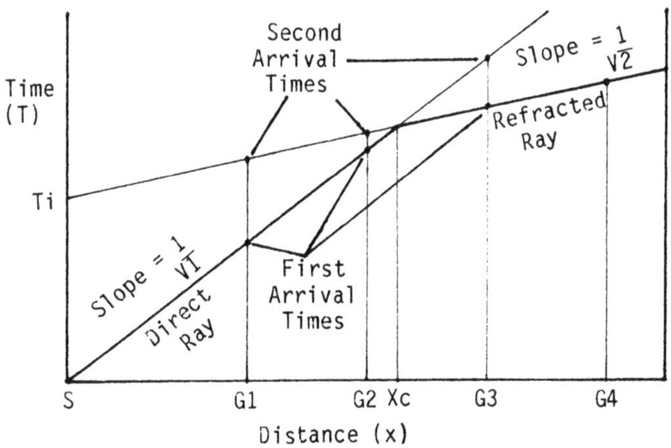

Figure 6.13: Time-Distance Graph for a Seismic Refraction Survey in a Two Layer Case.

can be shown that:

$$T_i = 2h\sqrt{\frac{1}{V_1^2} - \frac{1}{V_2^2}} \quad \text{.......................(6.20)}$$

or $$h = \frac{T_i}{2} \cdot \frac{1}{\sqrt{\frac{1}{V_1^2} - \frac{1}{V_2^2}}} \quad \text{.....................(6.21)}$$

and $$x_c = 2h\sqrt{\frac{V_2 + V_1}{V_2 - V_1}} \quad \text{.....................(6.22)}$$

or $$h = \frac{x_c}{2}\sqrt{\frac{V_2 - V_1}{V_2 + V_1}} \quad \text{.......................(6.23)}$$

Note that if there is little contrast between the two velocities it will be very difficult to determine the critical distance by Equation 6.23. It is usually better to determine h by means of the delay time method (Equation 6.24).

If there is a succession of beds with increasing velocity with depth, several boundaries may be detected and thus the thickness of the beds determined. If the boundary is dipping the method of survey and interpretation are more complicated.

The Seismic Reflection Method

This method is the most extensively used of all the geophysical prospecting techniques. As a result this section of applied geophysics is covered in great detail by most texts. It gives a more direct and detailed picture of subsurface structure than any other geophysical method. If the velocity of the wave is known then accurate determinations of depths to various boundaries can be made. The method is almost entirely used for petroleum prospecting.

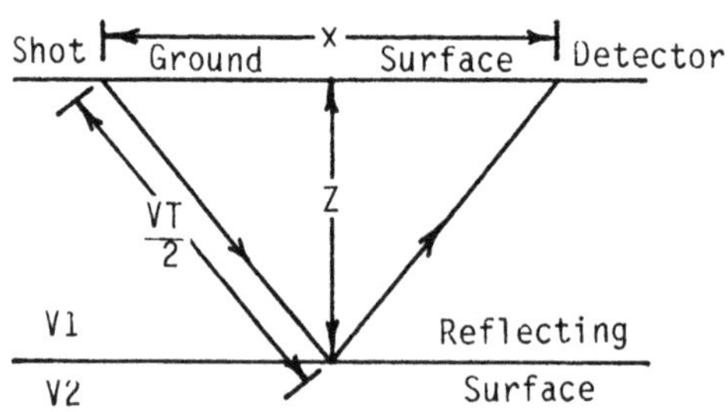

Figure 6.14: The Principles of the Seismic Reflection Method.

Figure 6.14 illustrates the path of that part of the energy of an incident wave which is reflected when the wave reaches the boundary between rocks of different velocities. A detector at the surface indicates the arrival of the reflected wave. If the velocities of the layers are known then the travel time can be used to calculate the depth to the reflecting surface. Otherwise, "depths" are referred to by their travel time and boundaries are called reflectors. The relevant equation is:

$$Z = \frac{1}{2}\sqrt{(\bar{V}T)2 - x^2} \quad \text{..............(6.24)}$$

where V is the velocity through the upper layer.

If there are a number of boundaries where successively deeper layers have increasing velocities then the waves fom successively deeper boundaries will arrive with successively increasing travel times at the detector.

The average velocity V may be determined by any of the following methods:

1. Well shooting where a shallow charge is fired near a deep well into which a detector has been lowered.
2. Continuous velocity logging where a tool approximately three metres long is lowered into a well. At one end of the tool is a sound source and at the other is a detector. A continuous log of velocity with depth is determined which is not only useful for interpreting seismic data but also for direct geological interpretation (see also the discussion of sonic logs in Chapter 9).

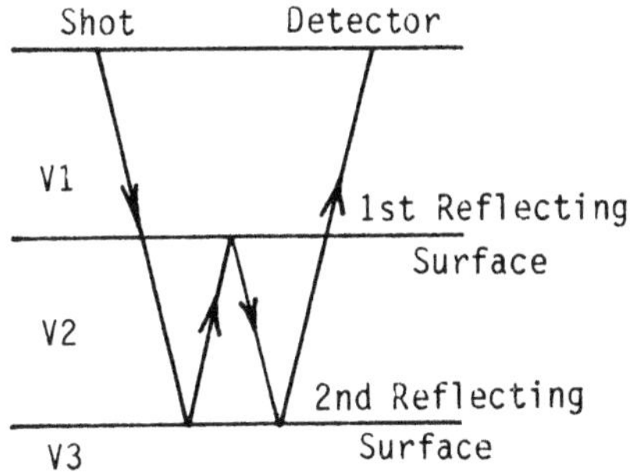

Figure 6.15: An Example of Multiple Reflections.

3. Analysis of reflection records themselves. If there are a number of good reflection horizons then by firing shots at varying distances from the detectors the average

velocities can be calculated.

During some surveys reflections are observed arriving so late on the records that they appear to have been reflected from a surface deep in the basement. However, it has been shown that these reflections have been reflected several times before being detected (Figure 6.15). The resulting traces on the profiles are called "ghosts".

ELECTRICAL

Introduction

There are many methods which may be used when carrying out electrical prospecting. They range from measuring natural electric fields (spontaneous-polarisation and telluric currents) to measuring the effects of introducing artificial currents into the Earth (equipotential, resistivity, induced polarisation and electro-magnetic methods). Electrical methods are mainly used in the search for metallic mineral deposits although some methods (resistivity in particular) are used in engineering and groundwater investigations. Electrical prospecting is little used in the initial stages of petroleum exploration mainly because of the limited depth of penetration (200-300 metres). However, electrical methods are used when logging oil wells when a great deal of information about various strata can be obtained (see Chapter 9).

Electrical Properties of Rocks

Resistivity. Resistivity, or conversely, conductivity, is the most important property of Earth materials when applying electrical methods. The relationship between resistance (in ohms, R), current (in amperes, I) and voltage (V) is given in Ohm's Law:

$$I = \frac{V}{R} \quad \ldots\ldots\ldots\ldots\ldots\ldots\ldots (6.25)$$

If a conducting cylinder of length l and cross sectional area S has a resistance R then the resistivity ρ is:

$$\rho = \frac{RS}{l} \quad \ldots\ldots\ldots\ldots\ldots\ldots\ldots (6.26)$$

The range of resistivities in rocks and minerals is very large. For example, materials with ionic conductivity (unconsolidated sediments to crystalline rocks) have resistivities of 10 to 10^5 ohm-metres; and materials with electronic conductivity (native metals to base metal sulfides) have resistivities of 10^{-7} to 1 ohm-metres.

The contrast of resistivities or conductivities between minerals of economic importance and their local environments is the basis of electrical methods. Field surveys measure apparent resistivities rather than laboratory determined resistivities. This results from influences of groundwater salinity and other factors.

Electrochemical Activity. This property depends on the chemical composition of the rocks and the composition of the electrolytes dissolved in the groundwater with which they are in contact. Electrochemical activity is most often used in the self-potential methods of exploration and well logging.

Dielectric Constant. This is similar to susceptibility in magnetism. It is a measure of the polarisability of a material in an electric field and is the basis of inductive prospecting techniques such as the induced polarisation method.

Current Flow in the Earth

The mathematics of electrical prospecting methods is considerably more complicated than for the gravity or magnetic methods. Also, mathematical interpretation of the results of surveys is less accurate because of a large number of variables. For this reason, the mathematical aspects of the following discussions is kept to a minimum.

When a current is introduced into the ground, a three dimensional current flow results (Figure 6.16). This assumes no variation in conductivity. The current density (the total current crossing any surface normal to the lines of current) is very high around the electrodes so the resistance around the electrodes is very high compared to an area far removed from the electrode. This is called the contact resistance of the electrode. This produces large potential differences in the vicinity of the input electrodes. To avoid this problem two additional electrodes are introduced at points where the potential gradient is negligible (Figure 6.16). To further limit any possible electrode potential with the detector electrodes, pot is used as the electrode. The contact is thus a liquid/liquid contact and no electrode potentials are developed.

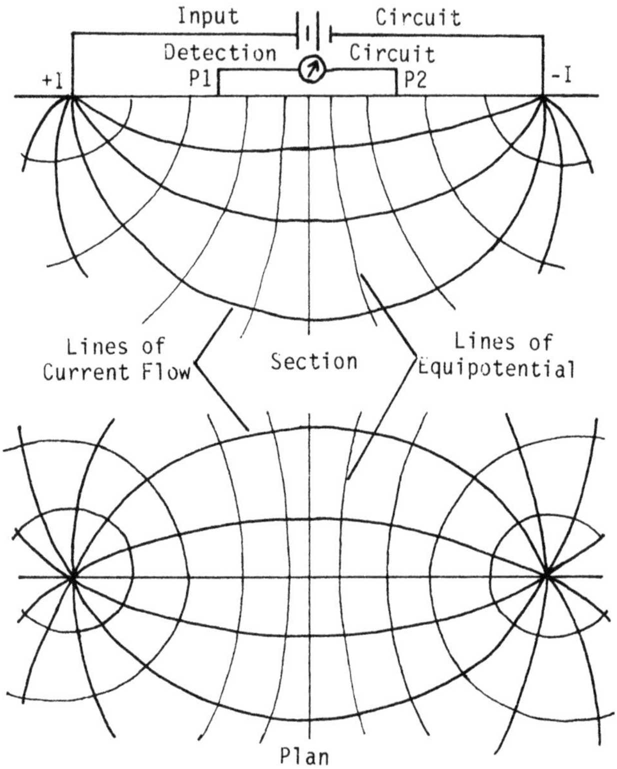

Figure 6.16: Lines of Current Flow and Equipotential in an Homogeneous Conductor.

Equipotential Line Method

If the conductivity of the ground is constant the ideal current flow is as shown by Figure 6.16. However, if a body with differing conductivity occurs in the surveyed area then the lines of current flow will be distorted as shown in Figure 6.17. Plotting the equipotential lines is relatively simple. Two porous pots are connected by a constant length of wire to a galvanometer and with one electrode fixed, the other is moved until the galvanometer registers a zero reading. Both electrodes must therefore be on a line of equal potential. If all such locations are plotted then a map of equal potential lines (and also lines of equal current flow) can be prepared. This can then be examined to see if any anomalously conductive material is present.

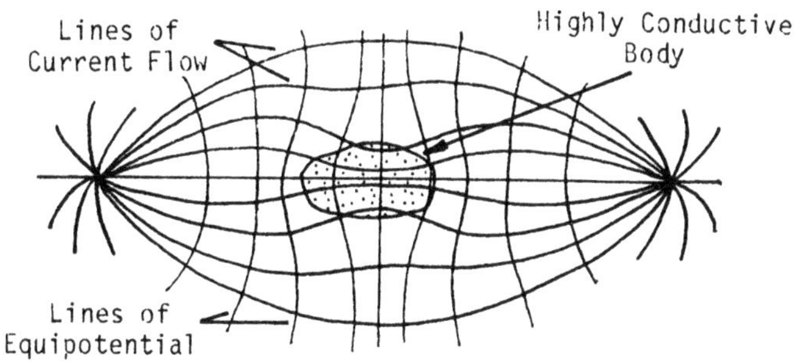

Figure 6.17: The Effect of a Body of High Conductivity on Lines of Current Flow and Equipotential.

Resistivity Method

Principles. By introducing direct currents into the Earth the apparent resistivities (ρ_γ) of Earth materials may be measured. The main factors which control the resistivity of rocks are the moisture content and the salinity of the ground water. This type of conduction where the charge moves as ions through waters is ionic conduction. In the case of some ore minerals, metallic conduction occurs, that is, the current movement is by means of electrons. Such conduction is of the order of seven times that of ionic conduction. However, the high conductivity of the electronic conductors is only likely to be met in massive mineralisation. There is little difference in the resistivities of disseminated mineralisation and the surrounding rocks.

The resistivity method is mainly used as a mapping tool to determine horizontal or vertical variations. The two main electrode configurations used are those of Wenner and Schlumberger.

The Wenner Configuration. The four electrodes are equidistant from each other as shown in Figure 6.18. If the electrode separation is progressively increased about the central point of the array then resistivities of material at increasing depths are measured (Figure 6.19). This is called the expanding electrode system or electric drilling. The close electrode spacing would measure

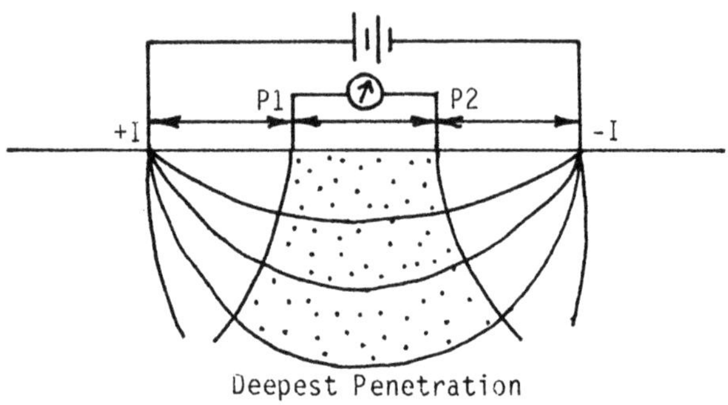

Figure 6.18: The Wenner Electrode Configuration.

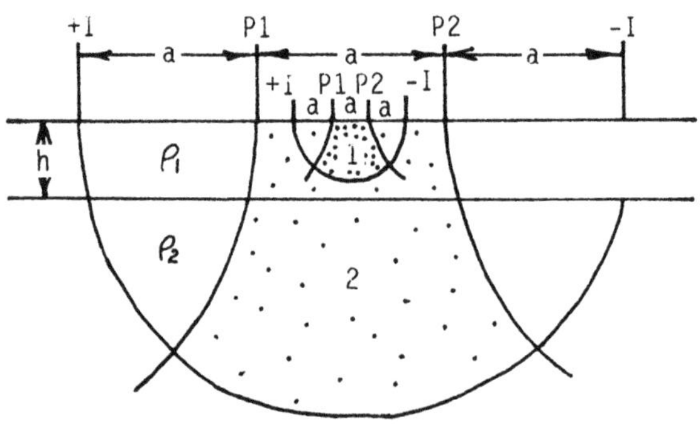

Figure 6.19: The Wenner Expanding Electrode System.

volume 1 and hence ρ1. The wider spacing would measure volume 2 and hence ρ2 for practical purposes. If a is plotted against ρ (Figure 6.20) then the depth (h) to the boundary between the two materials can be determined.

The Wenner configuration is also used to determine horizontal discontinuities. The electrode separation is kept constant but the array is bodily moved along a traverse line at increments equal to the electrode separation. This is called profiling.

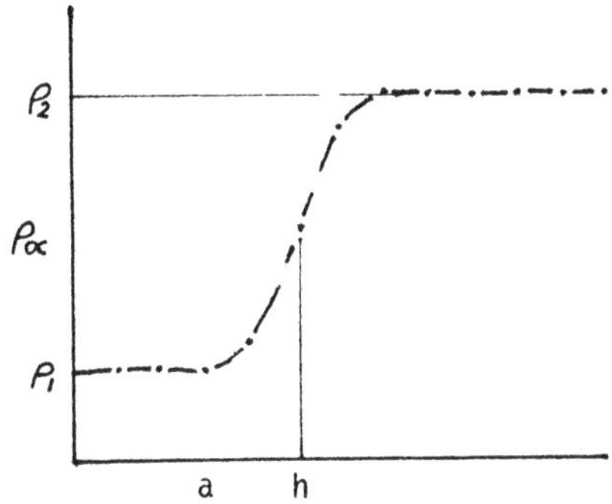

Figure 6.20: The Results of a Wenner Expanding Electrode Survey.

The Schlumberger Configuration. The detector electrodes are close together compared with the input

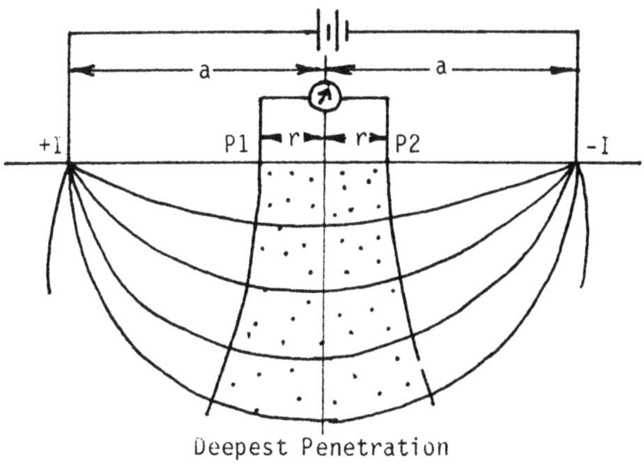

Figure 6.21: The Schlumberger Electrode Configuration.

electrodes (Figure 6.21). The measurements are not as influenced by horizontal variations as the Wenner configuration. Vertical discontinuities are measured by keeping the detector electrodes stationary but progressively moving the input electrodes further apart. Only a relatively thin but deep volume is measured so that vertical variations are insignificant.

Induced Polarisation Method

Disseminated mineralisation has little conductivity contrast with country rocks and the previously described electrical methods are generally unable to satisfactorily detect this type of mineralisation. The induced polarisation method was developed to overcome this problem.

If an electric current which is passing into the Earth through ground electrodes is interrupted, a residual potential can be measured through these or nearby electrodes for some time after the current is switched off. This is because the Earth materials have been polarised by the current. Polarisation is the separation of electric charges to form an effective diolar distribution of charge in an electric field.

Two types of polarisation in the Earth are:

1. Metallic polarisation which is used to detect metallic mineral deposits.
2. Non-metallic polarisation which is regarded as background "noise".

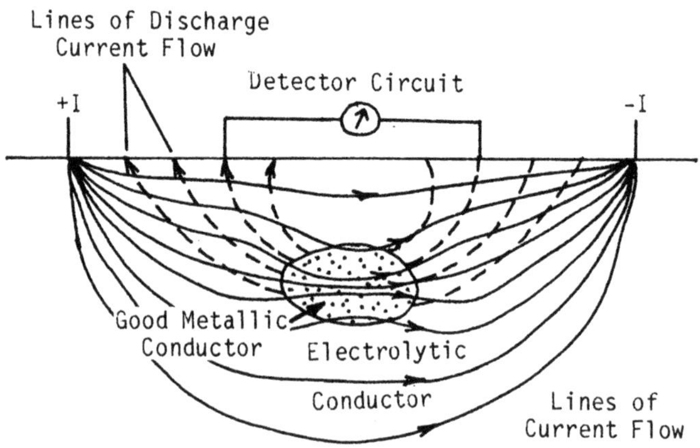

Figure 6.22: The Principles of the Induced Polarisation Method.

Polarisation occurs when a charge passes from an electrolytic conductor to an electronic conductor. For example, in Figure 6.22 accumulation of electropositive material occurs where the charging current passes from the electrolytic conductor and vice versa. This results in a dipole distribution which opposes the charging current. When the charging current is switched off the dipolar distribution of charge exist for a short time as it discharges itself as shown in Figure 6.22. Two potential electrodes are used to record the voltage against time and can be plotted as in Figure 6.23. The polarisation increases as the surface area increases with respect to volume. Therefore, the IP method is ideally suited to the search for disseminated mineralisation.

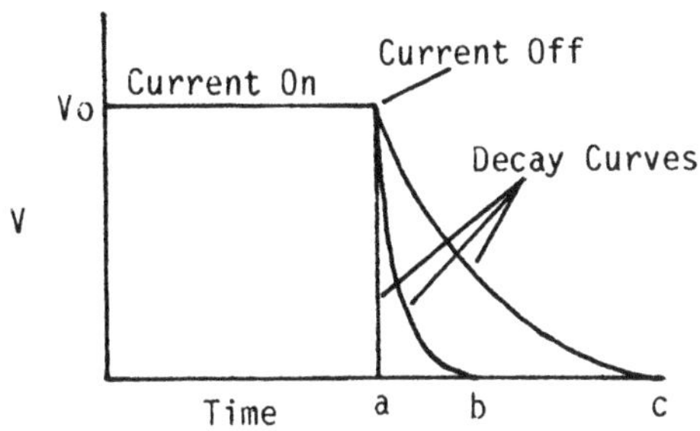

Figure 6.23: Decay Curves for Different Types of Polarisation.
(a) Instant Decay, no Polarisation.
(b) Rapid Decay, Little Polarisation.
(c) Slow Decay, Metallic Polarisation.

The non-metallic effect (or normal effect) is indistinguishable from the metallic polarisation effect as measured but is normally only one to two per cent of the metallic effect. The non-metallic effect can be dominant if clay particles occur in an electrolyte when ion exchange or electrodialysis occurs which is the same as metallic polarisation. The IP method is thus not very suitable in a highly saline and clayey environment. The two main methods of detecting the IP effect are described below.

Frequency Variation (or Frequency Domain) Method. An alternating current is passed between the input electrodes. The metallic particles behave as capacitors which are alternately charged and discharged. Resistance, conductance and capacitance all oppose the passage of alternating current, this combined opposition is known as impedance. Resistance is frequency independent but inductance and capacitance are frequency dependent so that when the frequency is changed so is the impedance. This method thus measures the IP effect for two different frequencies. The first, R1, at about 0.1 Hz and the other, R2, at about 10 Hz. A measure of the IP effect is the percentage frequency effect which is calculated from:

$$PFE = \frac{(R1 - R2)100}{R2} \quad \text{.....................(6.27)}$$

and a measure of both the IP response and the impedance is the metal factor (MF):

$$MF = \frac{PFE}{R1}(2\pi 10^3) \quad \text{.....................(6.28)}$$

where $2\pi 10^3$ is a function of the Wenner configuration. The metal factor emphasises the IP effects

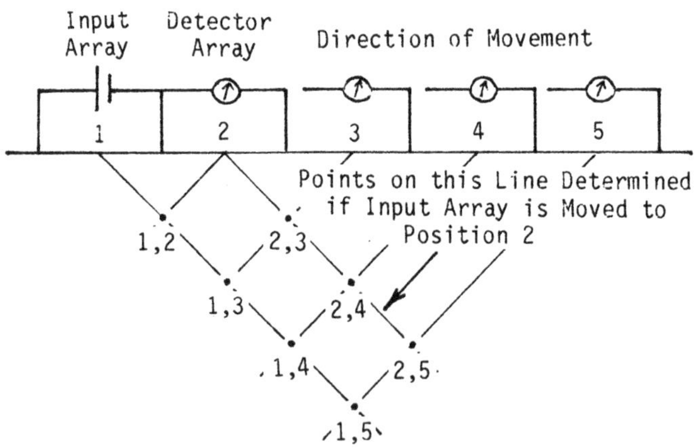

Figure 6.24: Dipole-Dipole Induced Polarisation Survey and Method of Plotting Results.

that occur in a conductive environment. This is because with a continuous conductor, R1 is small and the metal factor is high; and with a disseminated conductor, R2 is high and the metal factor is low.

<u>The Transient or Time Domain Method</u>. This is a

direct current method. The charging current is applied for approximately three seconds after which the decay of the IP is measured in the detector electrodes at a specific time during the decay (Figure 6.23). With more sophisticated equipment it is more common to integrate an area under the decay curve. The result is a millivolt/seconds factor. This is then divided by Vo (the polarisation voltage) to give milliseconds. Another factor used is chargeability, M, which is:

$$M = 1 - IP\% \qquad (6.29)$$

Both dipole-dipole and pole-dipole electrode configurations are used. The results are plotted as shown in Figure 6.24. The result is a purely qualitative representation of the locations of the factor plotted.

Electromagnetic

Electromagnetic methods are widely used during exploration for mineral deposits because of the advantage that no physical contact with the Earth is neccessary. Such surveys are commonly flown over frozen, arid, snow and even water covered areas.

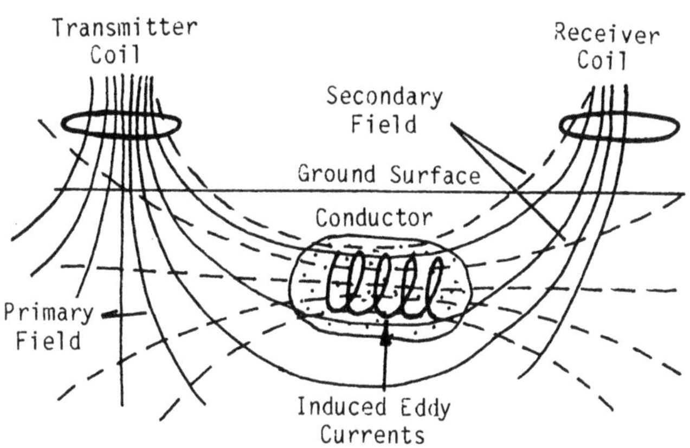

Figure 6.25: Principles of the Electromagnetic Method.

An alternating electromagnetic field (AMF) is produced by passing an alternating current through a coil of wire. The generated AMF waves spread out from the transmitting coil and if there is no subsurface conductor a receiving coil will detect this unaltered primary field. However, if there is a subsurface conductor, the primary AMF will induce currents by induction in the conductive body. These currents will then themselves generate a secondary AMF which will be detected by the receiver coil (Figure 6.25). The response is better than for other conductive electrical methods. Pyrrhotite, chalcopyrite and galena are good target minerals. Graphite and water-bearing fissures are troublesome because they may generate spurious AMFs. Second World War mine detectors and their offspring, the recreational metal detector, are examples of EM systems in which the transmitting and receiving coils are contained within the one instrument.

The receiver response is based on:

1. Change in the direction of the field
2. Change in the intensity of the field
3. Change in the phase of the field

The third response is the one which is used most often. When a primary field is applied to a conductor the induced eddy current is at right

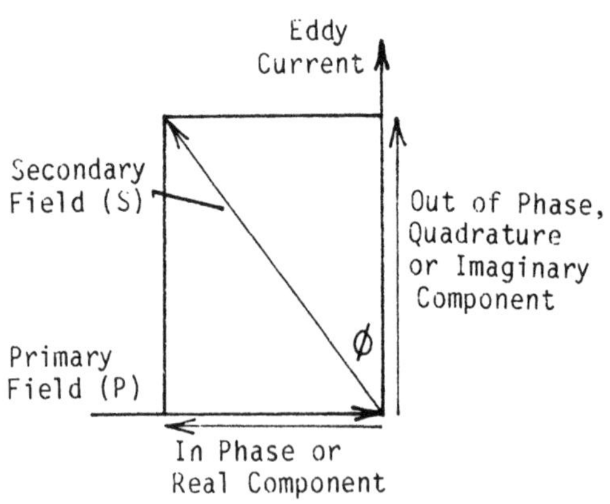

Figure 6.26: Phase Relationships in an EM System.

angles to the primary field. The secondary field as measured is the resultant of the primary and eddy currents (Figure 6.26). If the resistance of the body producing the eddy currents is large (that is, the conductivity is low) then ϕ approaches 0 and the imaginary component is large and the real component is small. The converse is obviously true. The real and imaginary components are usually expressed as percentages of the primary field. The example of an moderate conductor surveyed by a horizontal loop method where both the transmitter and the receiver move is shown in Figure 6.27.

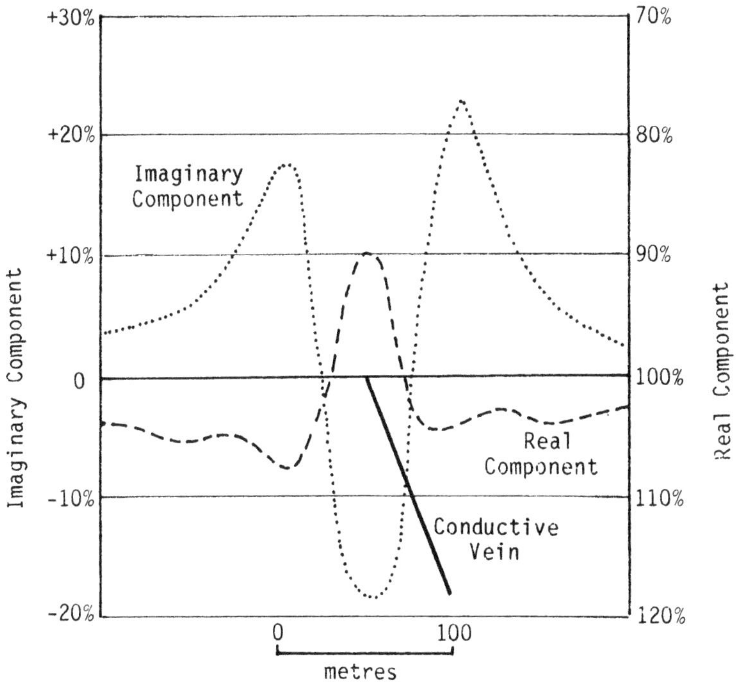

Figure 6.27: Results of an EM Survey Over a Steeply Dipping Conductive Vein.

The depth of penetration is inversely proportional to the frequency of the primary current. However, one of the main problems with this method is that a highly conductive overburden such as salt lakes can prevent penetration. Interpretation of EM results can be by mathematical analysis or by using reduced scale models.

Self Potential

The origin of the self-potential or spontaneous polarisation of sulfide ore bodies is a matter of debate. The explanation by Sato and Mooney (1960) is most widely accepted. If a continuous metallic conductor exists above and below the water table (Figure 6.28) then it is assumed that an electric current is produced by separate but simultaneous reduction of oxidising agents near the surface and oxidation of reducing agents at depth. The result is a current flow towards the top of the body with the corresponding electron flow to the bottom.

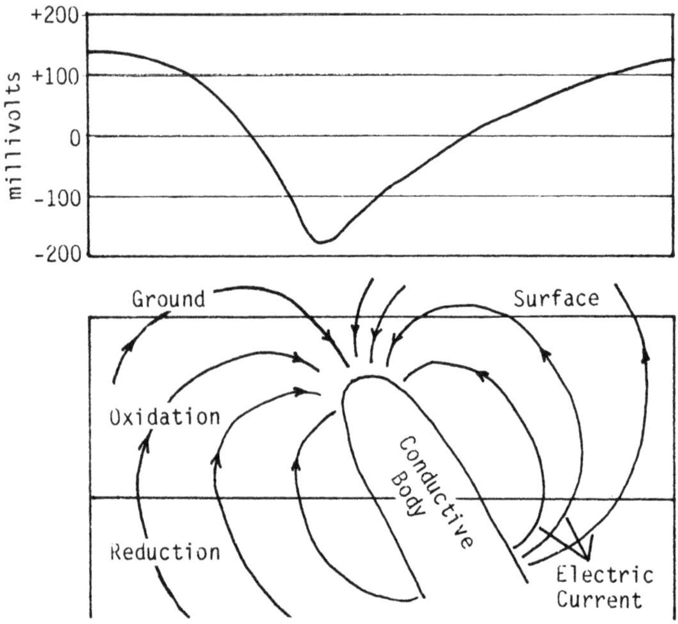

Figure 6.28: The Principles of the Spontaneous Potential Method and Results of a Survey.

The potentials are measured at the surface by a galvanometer and two porous pots. Note that this method is one of the few in which a negative anomaly is important.

RADIOMETRIC

Naturally radioactive materials such as the radioactive isotopes of uranium, thorium and potassium, produce three types of radiation:

1. Alpha Particles. These are helium nuclei and have a very short range (or maximum distance of travel) of only a few centimetres in air.
2. Beta Particles. These are electrons and positrons and although they have a much larger range then alpha particles, they are stopped by only a few centimetres of sand.
3. Gamma Rays. These are electromagnetic radiations with a theoretically infinite range. Unfortunately for mineral prospectors, these rays are stopped by one or two metres of rock or only a thin cover of soil.

Radiometric surveys detect gamma radiation, either by ground parties or by low-flying aircraft, using either Geiger counters or scintillometers. However, the limitations imposed by the lack of penetration of Earth materials by the radiometric particles mean that this technique is restricted in use.

REFERENCES

Dobrin, M. B., 1976. Introduction to Geophysical Prospecting (3rd Ed.). McGraw-Hill, New York.

Griffiths, D. H. and King, R. F., 1981. Applied Geophysics for Geologists and Engineers (2nd Ed.). Pergamon Press, Oxford.

Hanson, D. A., Heinrichs, W. E., Holmer, R. C., MacDougal, R. E., Rogers, G. R., Sumner, J. S. and Ward, S. H. (Eds.), 1966. Society of Exploration Geophysicists' Mining Geophysics, Volume 1, Case Histories. The Society of Exploration Geophysicists, Tulsa.

Parasnis, D. S., 1979. Principles of Applied Geophysics (3rd Ed.). Chapman and Hall. London.

Sato, M. and Mooney, H. M., 1960. The Electrochemical Mechanism of Sulfide Self-Potentials. Geophysics, **25**: 226-249.

FURTHER READING

JOURNALS

Geoexploration
Geophysical Prospecting
Geophysics
Mining Annual Review

Chapter Seven

EXPLORATION GEOCHEMISTRY

INTRODUCTION

Geochemistry is the science concerned with the chemistry of the Earth, in particular, the abundance, distribution and migration of the elements. Classic texts on "pure" Geochemistry (but which also refer to the economic applications of geochemistry) include Goldschmidt (1954), Rankama and Sahama (1950) and Krauskopf (1967).

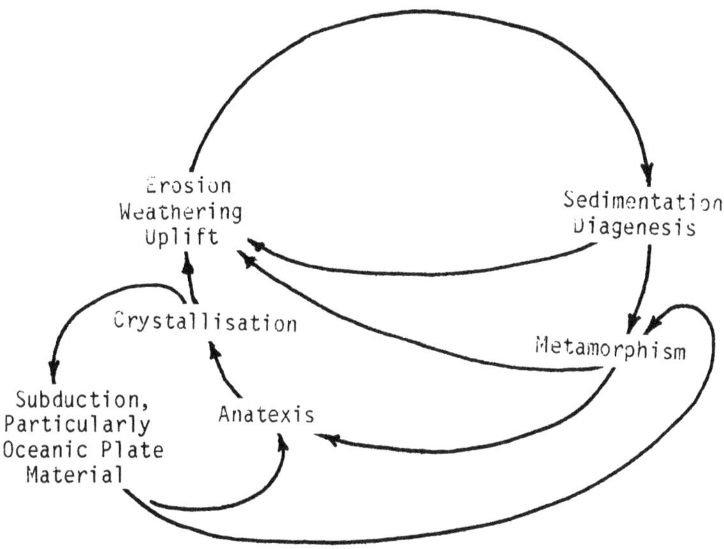

Figure 7.1: The Geochemical Cycle.

Exploration geochemistry is the application of geochemistry to mineral exploration. Although exploration techniques (such as gold panning) were applied by early mineral prospectors), modern exploration geochemistry developed in Russia and Scandinavia between the two World Wars. Geochemical techniques are also useful in environmental science. Standard texts include Rose, Hawkes and Webb (1979) and Levinson (1974). Other valuable references are the case histories presented at symposia, such as in Jones (1973).

MOBILITY

Mobility is the relative ease with which elements move in a geochemical environment which in turn is governed by the physical and chemical environment. For example, copper is more mobile in siliceous environments than in calcareous environments. The

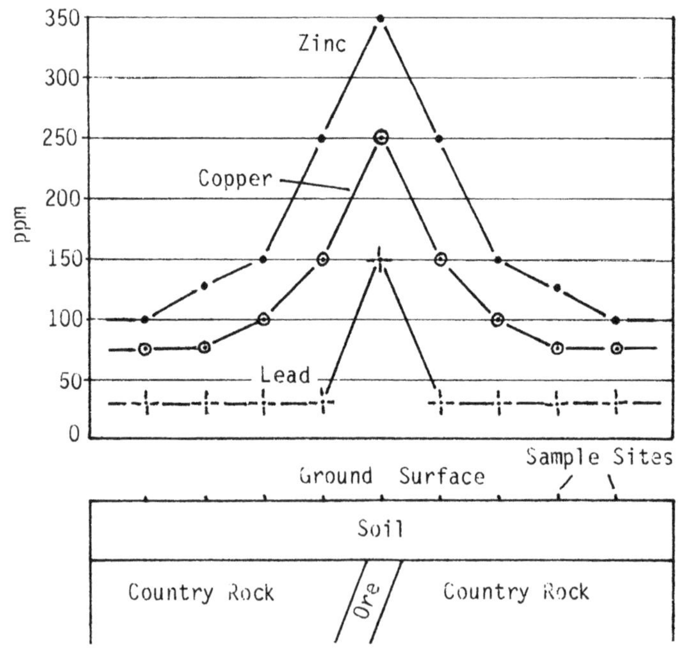

Figure 7.2: The Relative Degrees of Dispersion. of Zinc, Copper and Lead.

actual movement to, and deposition in, a new environment is called dispersion. Different elements have different mobilities and this factor is often used when planning geochemical exploration surveys. If the target is a copper-lead-zinc body then zinc could be used for regional or reconnaissance exploration because it is the most mobile and is therefore dispersed to a greater distance than the other elements. For increasingly detailed work copper and lead could be used in that order because of their relative decreasing mobilities (Figure 7.2).

In some cases it is more useful to analyse the samples for other than the desired element. This is because many mineral deposits contain several minor elements some of which are more readily dispersed, or removed from the site of mineralisation, because of their greater mobility. A classic example is arsenic in the presence of gold. Gold is relatively immobile and not appreciably dispersed whereas arsenic is very mobile. The arsenic associated with the mineralisation can be detected at considerable distances from the mineralisation. The arsenic is called a pathfinder element. Another example is the manganese halo that commonly surrounds volcanogenic exhalative base-metal deposits.

DISPERSION

Primary Dispersion

Primary dispersion occurs during the formation of a mineral deposit. It is usually associated with

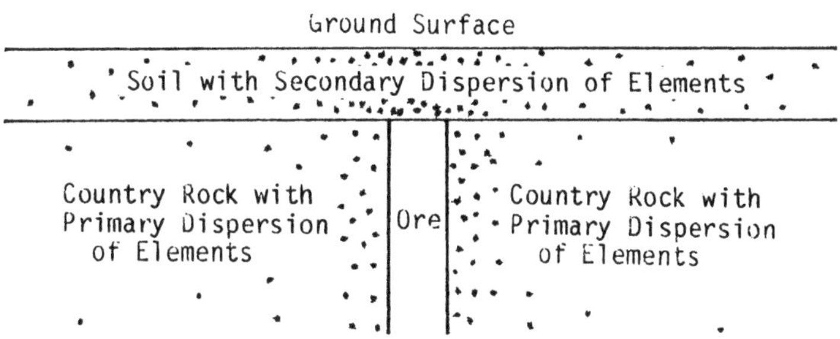

Figure 7.3: Primary and Secondary Dispersion.

deep-seated rocks when elements are diffused into the surrounding wall rocks of a deposit. This results in a low-grade halo of mineralisation which gives a larger, but weaker, target for exploration.

Secondary Dispersion

Secondary dispersion is dispersion which occurs after the formation of the mineralisation. The agents of secondary dispersion include metamorphism in deep-seated environments, and weathering in higher levels of the crust when the elements are redistributed in soils, surface waters and the air. This also results in larger targets for exploration (Figure 7.3).

GEOCHEMICAL STATISTICAL TERMINOLOGY

Provided a large enough sampling population exists the results of a geochemical survey may be interpreted statistically. Figure 7.4 illustrates the

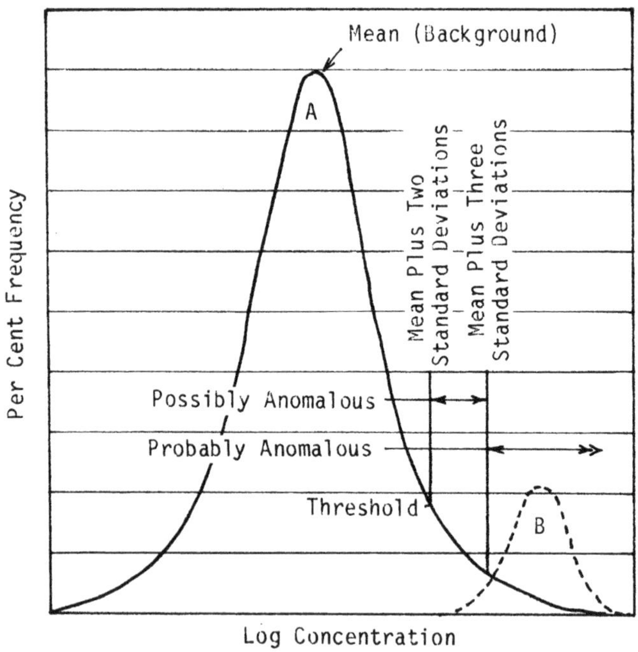

Figure 7.4: The Frequency Distribution of Two Populations of Geochemical Data.

frequency distribution of the results of a survey and the various parameters determined by statistical analysis. Note that the natural distribution of values in nature is log normal rather than normal.

Plotting the results in the above manner is useful in determining the limits of the various parameters in geochemistry such as background, threshold and anomalous values. The mean corresponds to the background and the mean plus two standard deviations is the value taken as the threshold. Values greater than the mean plus two standard deviations are considered to be possibly anomalous and values greater than the mean plus three standard deviations are considered to be probably anomalous. The inclusion of the values represented by curve B would result in a bi-modal distribution which is indicative of two separate populations. This second population may represent mineralisation. An alternative way to treat multiple populations is to plot a cumulative frequency curve.

Often, because of lack of sensitivity of the analytical method, the above ideal shape illustrated by curve A is not attained. Although such statistical methods are very valuable in interpreting geochemical results, especially when anomalies are very subtle or when several populations are present, a common rule of thumb is to consider any value two or more times the background value as anomalous.

Background

This is the "normal" abundance of an element in barren Earth material including rock, soil, water, stream sediment and air. Table 7.1 gives the median values of abundance of some of the more important elements in some rocks. Background values vary from locality to locality. In the case of secondary dispersed material the background values depend on many factors such as parental lithology, climate and topography. Large areas of the Earth's surface which have a consistency of geochemical characteristics are called geochemical provinces.

Anomaly

An anomaly is a deviation from the norm, or background, and although an anomaly can be either positive or negative the more valuable anomalies in mineral exploration are positive. However, not all positive anomalies indicative of target mineralisation. False anomalies can be produced by many factors such as preferential leaching by plants or adsorption of elements by iron/manganese oxides in reducing conditions.

Threshold

The threshold is the upper limit of normal background fluctuations and, therefore, the lower limit of positive anomalies. Two types of threshold are commonly recognised:

1. Regional thresholds which are the upper limit of the regional background values.
2. Local thresholds which are the upper limits of local background values. These local background values may be due to the weak mineralisation halo which often surrounds a mineral deposit due to primary dispersion.

The distinction between the two types of threshold is important when planning a geochemical survey. The aim of reconnaissance surveys should be to detect the plateau of values rising from the regional to the local threshold values. Detailed exploration can then be restricted to locating anomalies in these areas. Figure 7.5 illustrates

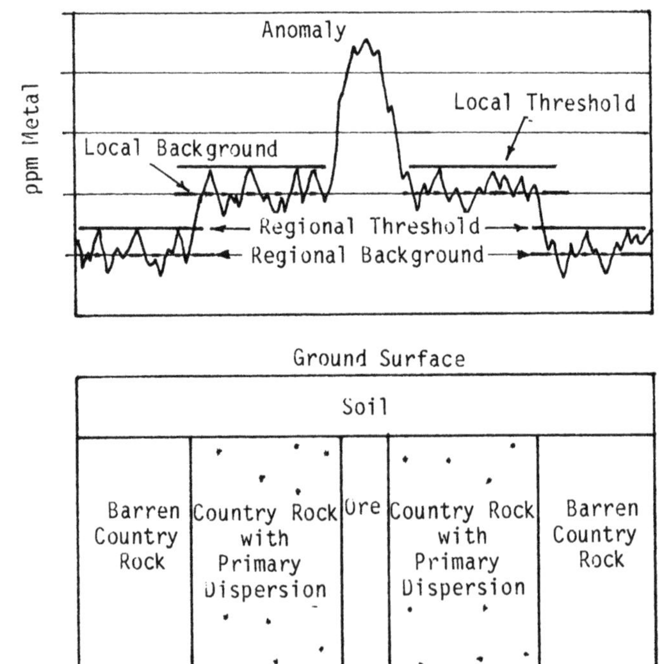

Figure 7.5: Regional and Local Background Values.

the relationships between these features. The variations in the values of the features determines the geochemical relief of an area.

Contrast

Contrast is the relative difference between the anomalous and the background values. Factors which affect the contrast are:

1. The primary contrast between the mineralisation and the country rock
2. The relative mobility of the elements
3. Dilution of anomalous values by barren material

As a result, the contrast in any area may vary with size fraction, soil horizon or extraction technique. It is one of the purposes of an orientation survey to determine which combination of the above will result in the highest contrast.

TYPES OF SURVEYS

Orientation Surveys

An orientation survey is carried out first to determine which combinations of techniques to use when planning a geochemical survey. The purpose of an orientation survey is to determine the optimum parameters which will give the best chance of locating mineralisation. These include the sampling density, the size fraction of the material to be analysed, the elements to be determined and the extraction technique. These aspects are more fully considered in the sections on the different types of geochemical surveys.

Overburden Surveys

Overburden is any material overlying bedrock including residual and transported soils. The size of the target determines the sampling density because,

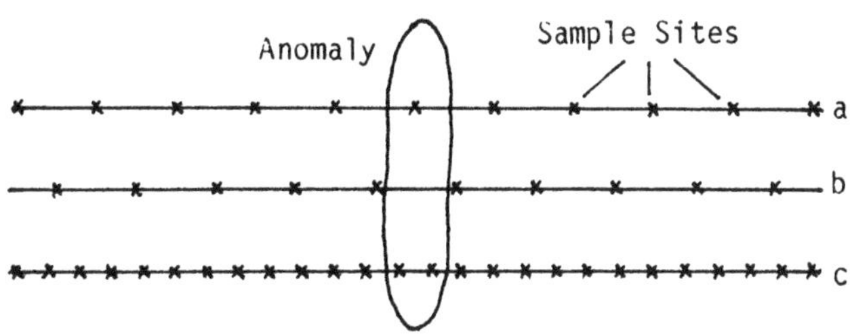

Figure 7.6: Size of Anomaly and Sampling Density.

as a rule of thumb, at least two samples should

be taken fom an anomalous area to ensure that an anomaly is not missed (Figure 7.6). If the sampling interval is as for lines a and b then the anomaly may not be detected because the interval is too great (as in line b). Line c indicates a suitable sampling density for such a target.

Sampling of different soil horizons can also give different results. This is because the normal soil forming processes result in elements being leached from the A horizon to the B horizon. Figure 7.7 illustrates this relationship in the transported soils of the Eastern Goldfields district of Western Australia. The reason that the soil does contain elements from the underlying mineralisation even though the soil is transported is because the vegetation in this arid climate has developed deep root systems. The elements are transported up through the plant to the leaves which then fall to add the elements to the humus layer of the soil. These elements are subsequently leached to the B horizon. This is a simple example of how an orientation survey can give results other than those expected by dogma.

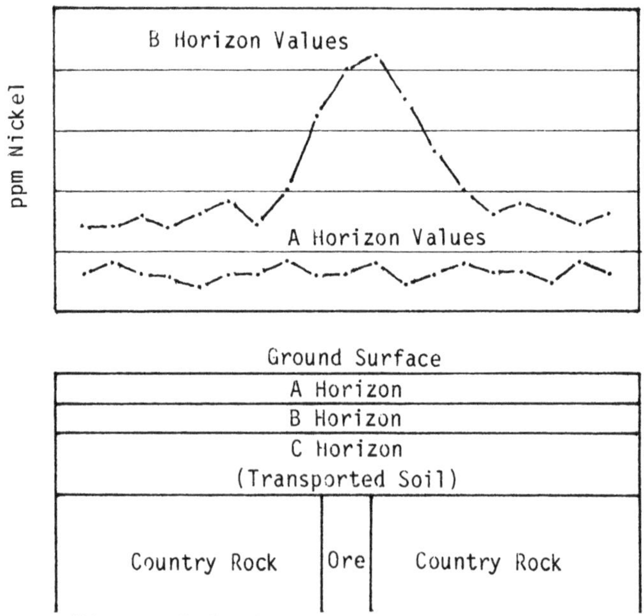

Figure 7.7: Geochemical Anomalies in Transported Soils.

Different size fractions can also give different results because the elements can be dispersed mechanically or chemically. If the dispersal is mechanical then it is more likely that the coarser fractions will give better contrast. If the dispersion is chemical then the silt to clay fraction would give better contrast because the elements tend to be adsorbed onto the clay particles.

Stream Sediment Surveys

Dispersion of elements is generally greater in stream sediments than in soils. As a result, areas with well developed drainage are suitable for reconnaissance surveys which are carried out with a low sampling density. Higher sampling density soil surveys are then carried out during the detailed exploration. In areas of poorly developed drainage, such as many parts of Australia and Africa, reconnaissance soil sampling is used exclusively.

The principle behind stream sediment surveys is that a sample of sediment represents an average of the material in the drainage basin upstream from the sampling point. The larger the stream the greater the drainage area. This results in less variation in the elemental concentration due to variation in lithology so a low density sampling programme is sufficient. The larger the streams, the lower the sampling density and the less detailed information is gathered. However, this method is valuable in determining geochemical provinces.

As the size of the stream sampled decreases so does the drainage area and hence there is more variation in the elemental concentration in the samples. This in turn means that the sampling density must be increased to make sure the variations are detected.

Sampling density is also determined by the length of the dispersion train. This is the downstream distance from the source of anomalous material where the elemental concentration is lowered to background values due to dilution from barren material washed into the stream (Figure 7.8).

The size fraction sampled is generally more critical in stream sediment surveying than in soil. This is because material is transported by both mechanical and chemical means. Mechanically transported material tends to be coarser than chemically transported material. Mechanical transportation tends to occur in areas of physical weathering and high relief such as the Flinders Ranges of

South Australia. Conversely, chemical transportation tends to occur in areas of chemical weathering and low relief such as many parts of central Africa.

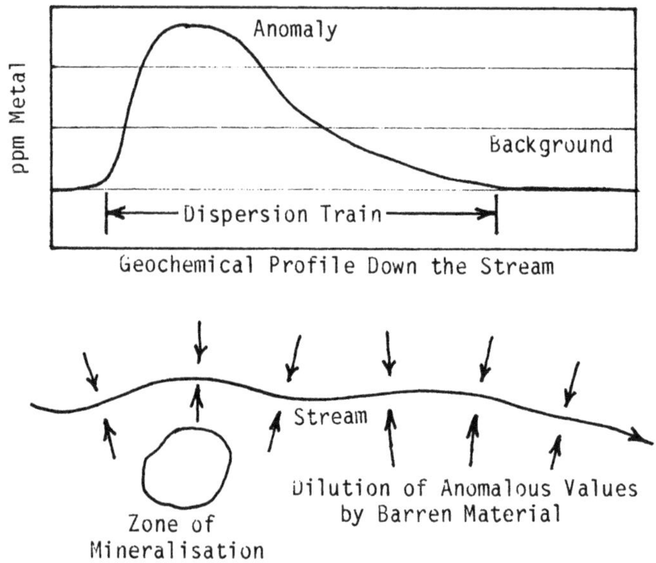

Figure 7.8: The Development of a Dispersion Train.

Vegetation Surveys

Vegetation surveys are becoming more popular. However, collection of the correct part of the plant and preparation of the samples presents some problems particularly in terms of time. Factors which influence concentration of elements in vegetation samples are:

1. The age of the part of the plant sampled. The most widely used is one year old growth
2. The portion of the plant (root, stem, leaves or flowers)
3. The climate both in gross terms and seasonal variations
4. The soil type which ultimately (in the case of residual soils) depends on the bedrock geology and any mineralisation

One of the advantages of a vegetation survey is that if suitable deep rooted species occur in an area covered by transported overburden the plant may be able to extract elements from the underlying bedrock. The elements then travel up the trunk through the branches to the leaves which are easily collected. The elemental content of the leaves may then reflect bedrock geology. In some cases the plants actually concentrate the elements which makes detection easier. Brooks (1973) covers this topic in great detail.

Rock Surveys

The most obvious rock surveys are the routine analyses of rocks during mining and drilling operations. Not only are the elemental concentrations of mineralisation determined but primary dispersion haloes around mineralisation can be detected. Routine sampling of outcrops during regional mapping can also outline geochemical provinces.

The biggest problem with rock sampling is the reliability of the sample taken as being indicative of what is generally a heterogeneous rock mass. The larger the grain size the more unrepresentative is the sample. If at all possible, soil or stream sediment sampling is preferable because of the averaging effect of weathering and soil/sediment formation.

Water Surveys

Although sampling water is simple in theory, in practice a big drawback is the problem of contamination of the sample once collected mainly because of the very low concentrations encountered. For example, most elements occur in concentrations of parts per billion (ppb) rather than parts per million (ppm). In addition, variations in pH and Eh can make interpretation of the results very difficult, especially when dealing with ground water. An additional problem is the weight of the sample and the difficulty of transporting it from the field site.

Vapour Surveys

This type of geochemical survey is a relatively new development. The two main types are air and soil-gas.

Mercury is the most common vapour detected during air surveys as mercury is associated with many types of sulfide mineralisation. Other vapours indicative of sulfide deposits are sulfur dioxide and hydrogen sulfide.

The most frequently used gas in soil-vapour surveys is radon. This gas is formed by the decay of radioactive minerals and as a result soil-vapour surveys are becoming routine in uranium exploration programmes.

There are a number of problems associated with vapour surveys, the most important being the very low concentrations of the vapour. Other factors apart from geology can also affect the concentrations of the vapour. These include barometric pressure, temperature and winds. Airborne vapour surveys generally detect mineral districts rather than individual deposits.

Isotope Surveys

These surveys determine the isotopic abundances of carbon, sulfur, oxygen and lead. The results can be expressed as simple ratios or they can be related to the ratio of isotopes in relevant standards.

The value of this technique is that the different abundances can suggest either the origin of the element or the temperature at which the host deposit formed. The geochemical value of sulfur and carbon isotopes is reviewed in Rye and Ohmoto (1974), and the value of lead isotopes is reviewed in Doe and Stacey (1974).

Isotopic variations in carbon can give an indication of the place of origin of the carbon. The origin may be simply organic, marine or deep seated.

A narrow range of sulfur isotope ratios indicates a magmatic source of sulfur and a wide range of values a biogenic source.

Oxygen isotope ratios are considered to be accurate indicators of the temperatures of formation of gangue minerals that accompany ore minerals.

Lead isotope ratios are used to solve problems of ore genesis (Doe and Stacey, 1974). For example, Mississippi-Valley type lead-zinc ore-bodies have a Pb206/Pb207 ratio of about 1.4 to 1.3 (the so-called "Joplin" leads) compared with "ordinary" leads with a ratio of 1.2 or less.

Presentation of Results

The results of a survey can be presented graphically. The most common method of visually presenting soil, rock or vegetation survey results is by plotting the results on a map and then contouring the values. In the case of stream sediment surveys the results can be indicated by circles of different radii representing different values at each sample site. Alternatively, a worm diagram can be constructed. In a worm diagram the thickness of the stream upstream from the sample site is in proportion to the concentration at that site (Figure 7.9).

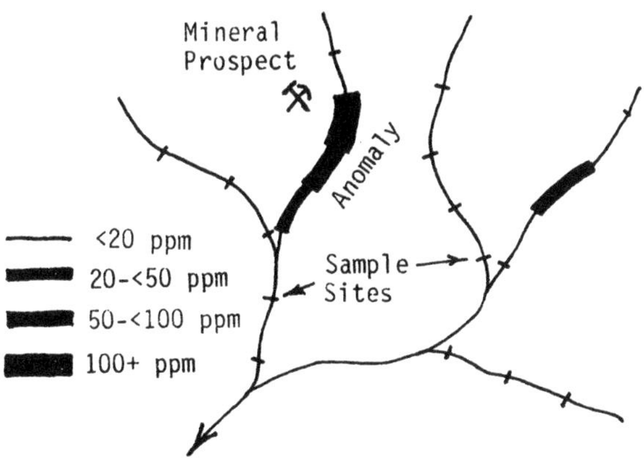

Figure 7.9: Representation of Stream Sediment Survey Results by a Worm Diagram.

It is often valuable to try to determine if there is a relationship between different elements taken from the same area. For example, in a number of deposits such as the White Pine copper deposit in North America and the Mt. Isa copper-lead-zinc deposit in Australia there is good correlation between organic carbon and metal content. Correlations such as these suggest that there may be a genetic relationship between the organic content of sedimentary rocks and base metal mineralisation.

ANALYTICAL METHODS

Preparation of Sample

Samples are dried and, in the case of soil and stream sediment samples, sieved into different size fractions (so that the maximum contrast can be obtained subsequently). Rock and vegetation samples are generally crushed and/or pulverised to make the sample homogenous.

Decomposition of Samples

Decomposition of the sample is necessary so that

the element to be determined can be isolated from the other constituents. Some of the more common methods of decomposition are:

1. Volatilisation where the sample is wholly or partly decomposed by means of an electric discharge or flame
2. Fusion where the sample is fused with an inorganic salt that melts at a low temperature and is capable of chemically attacking the sample. Some examples are $KHSO_4$, Na_2CO_3 and Na_2O_2
3. Acid attack by concentrated mineral acids such as HCl, HNO_3 and $HClO_4$; dilute mineral acids such as HCl and H_2SO_4; or solutions of complexing agents such as acetic acid and ammonium citrate
4. Oxidation by the ignition of vegetable matter

Separation

A common method of separation involves liquid/liquid solvent extraction. Two immiscible liquids, one being a suitable complexing agent, are added to the sample. The required element becomes preferentially soluble in one of the liquid phases. This technique is used mostly in colorimetric methods. Alternatively, the element(s) may be precipitated from solution.

Estimation

Accuracy and Precision. Different estimation techniques have different degrees of accuracy and precision. Accuracy may be considered an approach to the "real" value of concentration of an element. This is not as critical as precision in general geochemical exploration where the differences between anomalous and background values are important. Precision is a measure of reproducibility of results of a technique. Figure 7.10 illustrates the relationship between the "real" value of the concentration of an element in a sample and various degrees of accuracy and precision.

Gravimetry. The separated element is weighed. This technique is generally not sensitive enough for geochemical exploration.

Colorimetry. The dissolved or suspended constituent absorbs or scatters light of characteristic wavelengths. In many cases the degree of scattering or absorption is related to the amount of the constituent present and quantitative estimations of the constituent may be made.

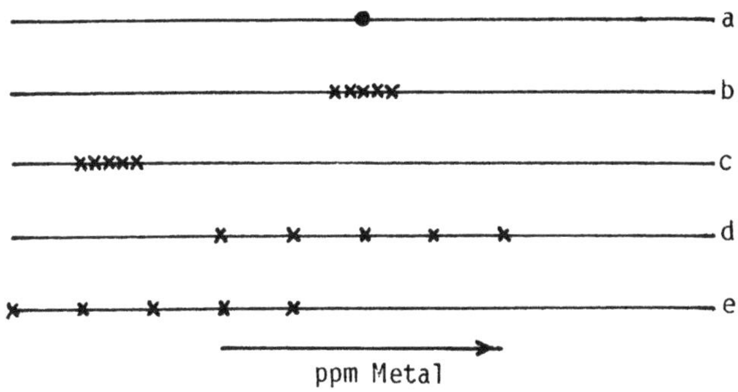

Figure 7.10: Accuracy and Precision.
(a) The "Real" Value.
(b) Accurate and Precise Determinations.
(c) Inaccurate and Precise Determinations.
(d) Accurate and Imprecise Determinations.
(e) Inaccurate and Imprecise Determinations.

Emission Spectroscopy. The sample is vaporised and radiation of wavelengths characteristic of the constituent elements is emitted. The wavelength and and the intensity of the radiation determines the element and its concentration respectively. Although the accuracy of the method is not high it has the advantage of allowing the simultaneous determination of a number of elements from the same sample.

Atomic Absorption Spectrophotometry. This technique enables rapid and inexpensive determination of a wide range of elements with good accuracy and precision. The sample is made into a solution which is then aspirated into a flame. A beam of appropriate radiation is passed through the flame. The degree of absorption is proportional to the concentration of the element. The machine may be calibrated using standards of known concentration allowing the concentration of the element can be determined.

REFERENCES

Brooks, R. R., 1972. Geobotany and Biogeochemistry in Mineral Exploration. Harper and Row, New York.

Doe, B. R. and Stacey, J. S., 1974. The Application of Lead Isotopes to the Problems of Ore Genesis and Ore Prospect Evaluation: A Review. Econ. Geol., **69**: 757-776.

Goldschmidt, V. M., 1954. Geochemistry. Oxford University Press, Oxford.

Jones, M. J. (Ed.), 1973. Geochemical Exploration 1972. Proceedings of the fourth International Geochemical Exploration Symposium, London, 1972. Inst. Mining and Metallurgy, London.

Krauskopf, K. B., 1967. Introduction to Geochemistry. McGraw-Hill, New York

Levinson, A. A., 1974. Introduction to Exploration Geochemistry. Applied Publishing Ltd., Calgary.

Rankama, K. and Sahama, T. G., 1950. Geochemistry. University of Chicago Press, Chicago.

Rose, A. W., Hawkes, H. E. and Webb, J. S., 1979. Geochemistry in Mineral Exploration (2nd Ed.). Academic Press, London.

Rye, R. O. and Ohmoto, H., 1974. Sulfur and Carbon Isotopes and Ore Genesis. Econ. Geol., **69**: 826-842.

FURTHER READING

TEXTS

Mason, B., 1960. Principles of Geochemistry. John Wiley, New York.

JOURNALS

Economic Geology
Geochimica et Cosmochimica Acta
Journal of Geochemical Exploration
Mining Magazine
Mining Annual Review

Transactions of the Institution of Mining and Metallurgy, London (Section B, Applied Earth Science)

TABLE 7.1

Approximate Median Abundances of Some Elements in Rocks (ppm)

Element	Igneous Rocks			Sedimentary Rocks		
	Granites	Mafic	Ultramafic	Limestones	Sandstones	Shales
Chromium	4	170	2980	11	35	90
Cobalt	1	48	110	0.1	0.3	19
Copper	12	72	42	5	10	42
Gold	0.0023	0.0032	0.0032	0.005	0.005	0.004
Iron	14,200	86,500	94,300	3800	9800	47,000
Lead	18	4	1	5	10	25
Nickel	4.5	130	2000	20	2	68
Silver	0.037	0.1	0.06	0.1	0.25	0.19
Sulfur	300	300	300	1200	240	2400
Tin	3	1.5	0.5	0	0.6	6
Uranium	3.9	0.53	0.03	2.2	1.7	3.7
Zinc	51	94	58	21	40	100

Data from Rose, Hawkes and Webb, 1979.

Chapter Eight

DRILLING TECHNIQUES

INTRODUCTION

By drilling holes into the Earth a geologist is able to determine the subsurface geology which may otherwise be inaccessible. In the case of ore deposits it is usually less expensive to obtain the added information by drilling rather than by shaft sinking or driving. Recent developments in technology allow drilling mine shafts and rises. Drill holes are also the main means of producing fluid Earth resources such as oil, gas and water. There are a number of different types of drilling techniques most of which have specialist uses.

Reedman (1979), Cummins and Given (1973), Peters (1978) and Giuliano (1981) contain chapters on drilling techniques. McGregor (1967) considers drilling as applied to mining geology. Petroleum drilling methods are described in Chilingarian and Vorabutr (1981) and Austin (1983).

PRINCIPLES

Drilling is usually carried out by crushing or shearing the rock at the bottom of a hole. Crushing (or percussion) methods include:

1. Cable (or Churn) Drilling
2. Hammer Drilling

Shearing methods break the rock by rotating a drilling tool, or bit, against the rock at the bottom of the hole. The different methods include:

1. Auger Drilling
2. Rotary Drilling
3. Diamond Drilling

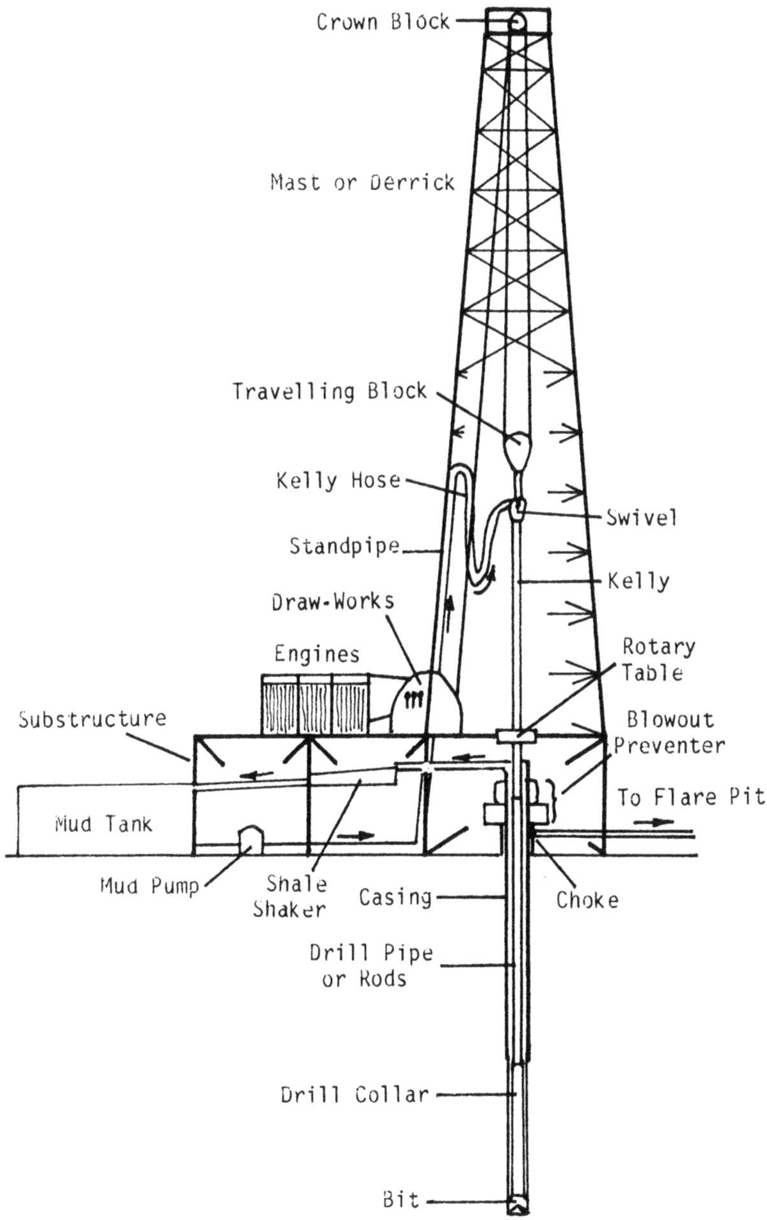

Figure 8.1: A Typical Rotary Drilling Rig.

Most drill rigs contain:

1. A power plant and a transmission system.
2. A system for raising and lowering the bit and associated drilling rods - the drill string.
3. A system for rotating the drill string (not required in a cable drill).
4. A system for circulating the drilling fluids (water, mud or air). Not required in a cable drill.

Figure 8.1 illustrates these components on a typical rotary drilling rig. Some specialist exploration drilling rigs are capable of rotary, hammer and diamond drilling.

CABLE OR CHURN DRILLS

These are relatively simple in operation but can obtain only limited information. Drilling is carried out by raising and dropping a heavy bit, or drilling tool, in a hole. The cutting edge of the bit is relatively blunt but it is made of hardened steel and the rock is crushed rather than chipped. It is necessary that the natural elasticity of the cable gives a "snap" to the bit to make the bit rebound after impact. The bit must operate in water so that the crushed material (or cuttings) is removed from the working face and to make easier the removal of the cuttings to the surface. This is done by substituting a bailer tool for the cutting tool at regular intervals - approximately after two metres have been drilled. Figure 8.2 illustrates a simple drill of this type.

Only vertical holes can be drilled in only relatively soft rocks or unconsolidated overburden. In either case it is almost invariably essential to support the walls of the hole to prevent caving, or collapse of the drilled material into the hole, and hence contamination of the sample. This support is provided by steel tube casing which is slightly larger in diameter than the tools. One string (or continuous length) of casing is put down the hole (or "run") until stopped by friction on the sides of the casing or some obstruction at the toe (or cutting edge of the casing). After further drilling a slightly smaller diameter casing is run inside the first to a greater depth.

Economics of the operation results in obtaining samples of crushed rock from an approximately two metre interval. This may be contaminated by material falling from higher levels in an uncased

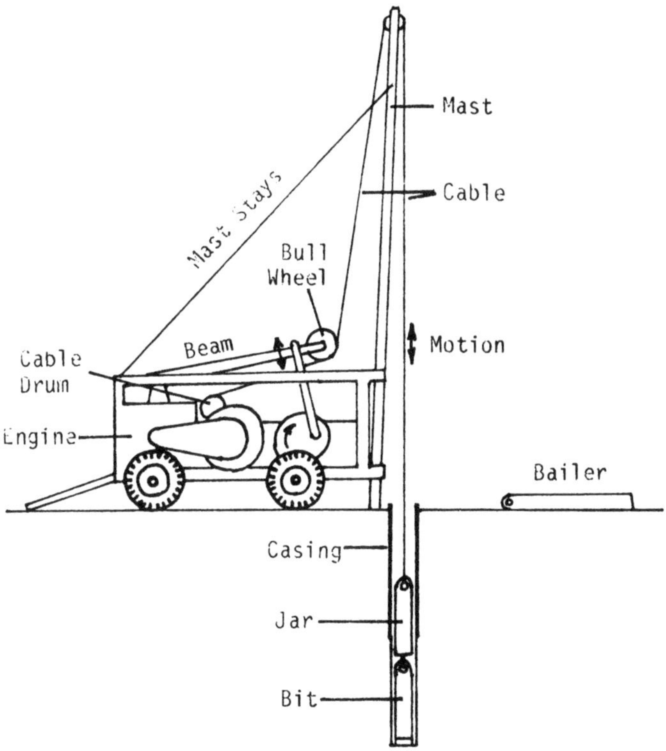

Figure 8.2: A Cable Tool Drilling Rig.

hole. Heavy minerals are also difficult to recover by the essentially simple bailer. However, this type of drilling rig can give useful information about the depth of unconsolidated material, soft rock successions and samples of ore. It is still widely used as a water drilling method and in the past was successful as a blast hole drill in open cut mining operations.

HAMMER DRILL

This is a modern adaption of the cable or churn drill. A pneumatic hammer, very similar to those used in road works, is attached to a drill string which is a number of drill rods which are screwed together. Although some hammer drills are fitted with the hammer at the top of the drill string, most have the hammer fitted at the bottom of the drill string - the down-the-hole hammer drills. The drill rods are made of high strength steel and are hollow.

Compressed air is forced down the centre of the drill stem and operates the hammer. The spent air then rises up the hole outside the drill stem carrying with it the rock particles dislodged by the hammer. At the surface, the air-dust mixture is passed into a cyclone which recovers the rock particles for examination or assay. An example of a small crawler mounted rig of this type used in mineral exploration is illustrated in Figure 8.3.

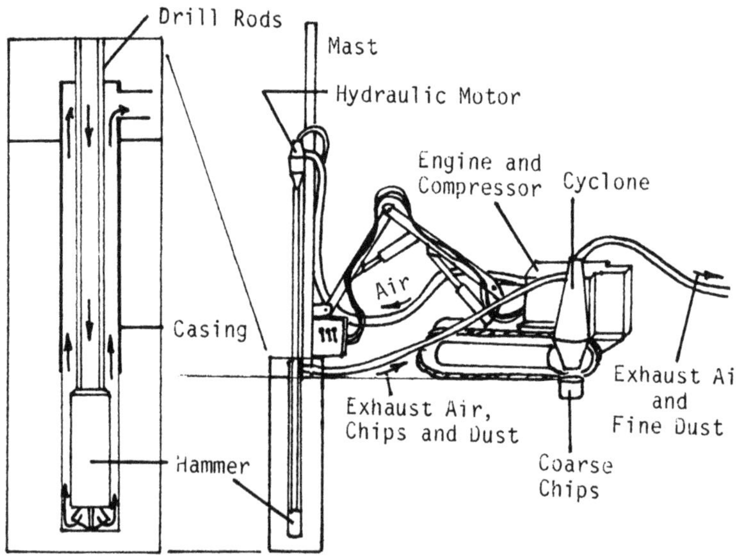

Figure 8.3: A Self-Propelled Crawler Percussion Drill.

This system of drilling works best in dry ground. Holes can be drilled at angles other than vertical although very flat holes present problems with dust recovery. Hammer drilling is used mainly for obtaining analytical samples and for drilling blast holes at open cut mines.

AUGER DRILLING

Auger drilling is used mainly for soil sampling and drilling holes for building foundation piers. The drill can be an auger bit, similar to that used in woodwork, which is connected to normal drill rods. More commonly, the drill rods are helical "flights" (Figure 8.4). These types of drills are capable of

Figure 8.4: A Tractor-Mounted Auger Drill.

drilling relatively unconsolidated sediment, and "soils" in an engineering sense. The whole drill stem must be removed from the hole periodically to retrieve the sample and to prevent clogging of the bit. This method, therefore, is used to drill only relatively shallow holes.

ROTARY DRILLING

There are a number of drilling methods in which the

drill string is rotated continuously during drilling operations. Hammer drilling, auger drilling and diamond drilling all rotate the drill string. However, the traditional rotary rigs are those used in petroleum exploration and development and in which the rotary action is applied by a rotary table (Figure 8.1).

The cutting tool commonly consists of three toothed cones (a tri-cone bit) which rotates as the drill string rotates grinding the rock as it does so. Immediately above the bit is at least one drill collar. This is a heavier than normal length of drill rod. The extra weight assists the bit in its cutting action and also tends to keep the hole vertical.

The broken-off pieces of rock, or cuttings, are brought to the surface by drilling mud which is pumped down the centre of the drill string. The cuttings are separated from the mud by a shale shaker which is simply a vibrating screen. The mud also lubricates and cools the bit.

The composition of the mud is usually a mixture of baryte, bentonite-type clays and other chemical additives. Such a mud can seal off permeable strata as the drill passes through them and stop loss of circulation of the drilling mud. The density of the mud also helps in containing any down hole high pressure which may be the result of high formation pressures caused by water or petroleum. This reduces the chance of a "blowout" - the uncontrolled movement to the surface of formation fluids under pressure. All rotary drilling rigs have a blowout preventer (BOP) fitted to the top of the drill casing. The BOP can seal the hole when the drill string is either in the hole or out of it.

A number of different drilling rigs have been developed for offshore work:

1. Drill ships are usually normal ships fitted with a drilling rig.
2. Semisubmersible rigs float on large pontoons. When these rigs have been towed into position the pontoons are partially flooded. The main mass of the rig is thus below wave action and results in a stable drilling platform.
3. Submersible rigs are similar to the semi-submersibles but rest on the sea floor after flooding.
4. Jack-Up rigs have three legs which are

raised when the rig is being towed to the drilling site. When the rig is in position, the legs are lowered to the sea floor and the drilling platform is jacked-up out of the reach of high seas.

Drilling at sea poses more problems than on land and is much more expensive. The well-head with a fitted BOP is on the sea floor. Connecting the sub-sea well-head to the drilling rig is a riser. This is a large diameter tube which has telescopic joints to allow for variations in depth due to tides and wave action. Risers have a similar function to that of normal casing. The drill string passes down through the riser and the returning drilling fluids pass upwards between the drill string and the riser.

When the target horizon is penetrated (or at other times when petroleum is encountered) a drill-stem test is usually carried out. This involves running the drill string into the hole with a test tool in place of the bit. The tool consists of a perforated tube above which are a rubber packer, a pressure recorder and a test valve. When the drill string rests on the bottom of the hole, the weight of the drill string forces the rubber packer against the walls of the hole isolating the perforated tube from the overlying mud column. Partially rotating the drill string opens the test valve and the formation fluids flow through the tool, up the drill pipe, through the choke to a burn pit. The pressure recorder produces a permanent record of the pressure information obtained.

DIAMOND DRILLING

Diamond drilling is probably tne most widely used drilling technique in mineral exploration. This is because, given the right ground conditions, a continuous core of rock may be obtained for the entire length of the hole. Recovery of the core assists in accurately determining rock types, boundaries and structures. In addition, any sample of mineralisation is representative of the interval being assayed.

Diamond drilling is a specialised rotary drilling technique. However, in distinction from the rotary rig described above, the rotary motion is applied to the drill string by a chuck which is identical in operation to the do-it-yourself electric drill. The cutting tool, or bit, is a tube of metal into which are set industrial diamonds. When the drill is in operation, the bit cuts a ring of rock and leaves a central cylinder or core. Immediately above the bit is a core-barrel. As the drill progresses, the core-barrel slides over the core. The core is prevented from falling out of the core-barrel by spring loaded jaws called core catchers (Figure 8.5). When the core-barrel is full of core the drill string is withdrawn, the core-barrel opened and the core removed.

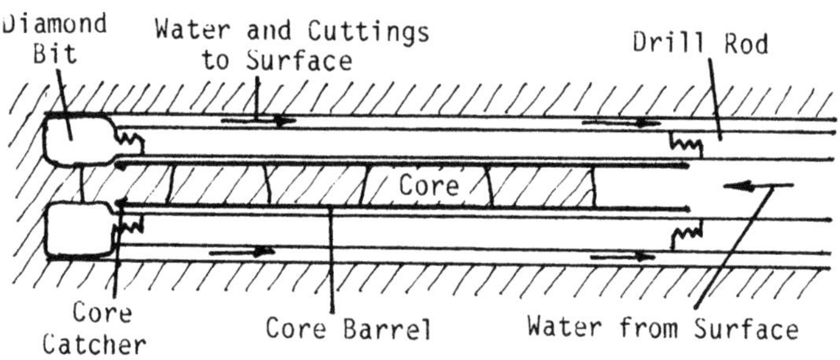

Figure 8.5: Diamond Drill Coring.

The removal of the drill string (to recover the core from the core-barrel, to replace a worn-out bit or to run casing) is a very time consuming operation. The drill string is raised by a distance equal to the length of individual drill rods. After the drill string is wedged in the hole, the top-most drill rod is unscrewed and stacked to one side of the rig. The drill string is then raised a similar distance and the process repeated. If the drill rods are short and/or the hole is deep, the time taken to retrieve the whole length of the drill string is often measured in hours. The process must be repeated when re-entering the hole.

A relatively recent advancement on this method is the development of wire-line drilling. In this method the core-barrel is smaller than normal and fits inside the drill rods. The core-barrel slides down the inside of the drill string and receives the core during drilling operations. When the core-barrel is full, the drill string is broken at the surface and a catching device called an overshot is dropped down the hole. This clips onto the core-barrel and the two are brought to the surface by a wire attached to a winch. The core-barrel can therefore be brought to the surface at any time without bringing up the whole drill string.

The diamond bits must be cooled and lubricated by water as in the rotary method. Figure 8.6 shows the field testing of a diamond drill core-barrel and bit. The water is pumped down the inside of the drilling rods, inside the core-barrel and through the drill bit. During drilling operations, this water returns to the surface up the outside of the drill string. As this return water rises to the surface it carries any rock particles dislodged by the drill bit. One bonus is that the cuttings brought to the surface can also be used for analytical or other work.

Figure 8.6: Field Testing A Diamond Drill Core-Barrel.

Diamond drilling can be carried out at any angle, including upwards, so that it is very useful during exploratory work in underground operations.

Diamond bits can also be attached to petroleum rotary rigs so that cores can be recovered and porosity and permeability measurements made.

As with any drilling method where a return of water, mud or air is an essential part of the operation, casing is almost invariably run in the first part of the hole where there is unconsolidated overburden or weathered rock. This initial part of the hole is often drilled by simple rotary or percussion drilling. Casing is also run when circulation problems arise such as when the drill enters fractured rock.

A relatively new technique called reverse circulation (or reverse flow or dual tube) drilling overcomes many of the problems when drilling in "bad ground". The drill rods are made up of an inner and an outer tube. The circulating fluid passes down the space between the two tubes, lubricates the bit and returns to the surface up the inside of the inner tube. The outer tube therefore acts as a casing which is always keeping pace with the bit. This type of drilling, using both tri-cone (or roller) and diamond bits, is becoming popular in many parts of Australia which have extensive deeply weathered rocks.

REFERENCES

Austin, E. H., 1983. Drilling Engineering Handbook. International Human Resources Development Corporation, Boston.

Chilingarian, G. V. and Vorabutr, P., 1981. Drilling and Drilling Fluids - Developments in Petroleum Science, v 11. Elsevier, Amsterdam.

Cummins, A. B. and Given, I. A. (Eds.), 1973. SME Mining Engineering Handbook (2 vols). Society of Mining Engineers of the American Institute of Mining, Metallurgical, and Petroleum Engineers, Inc., New York.

Giuliano, F. A. (Ed.), 1981. Introduction to Oil and Gas Technology (2nd Ed.). International Human Resources Development Corporation, Boston.

McGregor, K., 1967. The Drilling of Rock. CR Books, London.

Peters, W. C., 1978. Exploration and Mining Geology. John Wiley, New York.

Reedman, J. H., 1979. Techniques in Mineral Exploration. Applied Science Publishers, Barking.

FURTHER READING

JOURNALS

Geodrilling
Mining Journal
Mining Magazine
Mining Annual Review

Chapter Nine

BORE HOLE LOGGING

INTRODUCTION

The recovery of Earth materials by drilling is costly. Therefore, the maximum amount of information must be recovered and recorded. The main logging techniques are geological, geochemical, geophysical and mechanical. Individually or in various combinations, these techniques provide information vital for mineral and petroleum exploration and for engineering investigations. The material recovered may be consolidated or unconsolidated and may be in the form of fragments (cuttings) or various types of core.

Glenn and Hohmann (1981) present a brief overall review of well logging and borehole geophysics. Sections on this topic are also included in standard exploration geophysical texts such as Griffiths and King (1981) and Parasnis (1979). More detailed discussions of recent advances in this field are in Fitch (1982).

GEOLOGICAL LOGGING

Introduction

Any geological log should contain the following information as a preamble to the record of the logging as such:

1. Name of hole
2. Location of hole
3. Completion depth
4. Angle relative to the horizontal
5. Direction (if angle is less than 90 degrees)
6. For whom the hole is being drilled
7. Drilling company

8. Type of drilling rig
9. The name(s) of the driller(s)
10. Dates of start and finish of drilling
11. Name of the logger(s)
12. Type of sample being logged

Most exploration organisations have developed logging sheets for specific types of drilling and/or to suit their particular requirements. Some of these are machine readable and the information can be readily processed by a computer.

Different organisations also develop different logging techniques. However, it is advisable to log in the greatest detail possible in the time available. Summary logs can be prepared from detailed logs at leisure, but the process is not reversible in the cases where samples are no longer available. During logging, it is important to constantly check with the driller the interval being drilled. Another useful tip, and invaluable insurance, is to make carbon copies of log sheets.

Cuttings

Cuttings may be the result of auger drilling of unconsolidated material or percussion or rotary drilling of consolidated material. If consolidated, a representative sample of a bulk sample from an interval of drilling is washed to remove dust or drilling mud and then dried. If the material is unconsolidated, for example, from an auger hole in soils, then the samples are described as recovered as any treatment would alter the soil characteristics. A preliminary examination with a hand lens is followed by a more detailed examination using a binocular microscope.

The intervals described in the log are usually predetermined and are a function of the length of the drilling rods, for example, every one metre or every five metres. Normal geological parameters are described such as colour, grain size, texture, mineralogy, lithology, fossil content. Actual estimations of depth may be inaccurate because of the time taken for the sample to reach the surface. The difference between the true depth from which the sample came and the actual depth of the drilling bit when the sample reaches the surface is called the "lag". This is usually only a serious problem in very deep rotary holes drilled during petroleum exploration and development. Corrections can be made for lag when wire-line logs are run and accurate determinations of geological boundaries are made.

An effective visual representation of the log of a hole can be made by sprinkling some of the cuttings onto a glue covered wood or plastic strip. The strip is divided into even intervals corresponding to scaled sample intervals. The strip can be used to illustrate the lithology of the hole directly or it can be colour photographed for multiple distribution or inclusion in reports or papers.

Core

Unconsolidated core material, such as that recovered from drilling of soils and modern sediments, is usually protected from dessication by enclosing the core in plastic and then storing in core trays. When logging begins, the core is usually bisected with a knife and the cut surface, in particular, is logged.

Consolidated core, such as diamond drill core, is laid out in core trays as soon as it is removed from the core barrel. It is very important to mark the start (top) of the string of core in the tray and the depth interval from which each core barrel length came. A wooden block should also be placed at the end of each "run" of core. These details assist the geologist during logging and also makes easy the calculations of core recovery. Core recovery is determined by measuring the length of the core recovered and expressing that as a percentage of the corresponding drilled length.

Core is more valuable than cuttings because:

1. Accurate determinations of depths can be made
2. Details of relationships between different lithologies and structures can be observed
3. Attitudes of planar features can be determined (usually expressed as an angle with the core axis)
4. Undisturbed samples can be tested for physical properties
5. More representative samples are obtained for analysis or metallurgical testing

After logging in the field is complete, the core is taken to a core library or laboratory where further work may be carried out. It is normal to saw diamond drill core into halves or quarters so that different techniques can be applied to the same interval of core. It is in the laboratory that a number of different chemical staining techniques may be applied. These assist in identifying different carbonate, feldspar and sulfide minerals. Mineral exploration companies normally send a quarter of a core to a mineralogical or analytical laboratory for geochemical analysis.

There are a number of non-destructive methods of representing core material in a very realistic manner. As with the case of cuttings, colour photography is widely used. Methods which actually incorporate minute quantities of the rocks and minerals are the acetate peel and streak-print techniques.

Acetate peels of the sawn and polished flat surfaces of the core are made by pouring a thin layer of fluid acetate onto the surface. When dry, the acetate is peeled off, taking with it a very thin layer of the rock. The peels can then be mounted onto glass slides and examined using a microscope (Allman and Lawrence, 1972).

Alternatively, the streak-print technique relies on the removal of the fine mineral powder by adhesive tape. The tape is then mounted onto a white background, such as card, to reproduce the streak colours and textures of the rock (Morris and Ewers, 1978).

Portions of the core may be removed for thin section and polished section petrographic, mineragraphic and palaeontological investigations. Any material removed should be replaced by a wooden block containing details of the interval, identification number of the sample and the type of work done.

GEOCHEMICAL LOGGING

Mineral Exploration

Representative samples of cuttings or core are usually analysed for a range of elements the choice of which is determined by the exploration target of the company. However, regular scans for a greater range of elements should be carried out - you don't find what you don't look for! Samples may be analysed for:

1. A range of elements by the cheap but relatively inaccurate emission spectroscopy
2. A select few elements by the more expensive but more accurate atomic absorption spectrophotometry
3. One or more components, either elemental or

compound, by standard chemical techniques

Results are tabulated and also plotted as histograms or profiles along the plotted position of the drill hole.

Petroleum Exploration

Cuttings are analysed for organic carbon and carbonate carbon to determine potential source rocks. Polished sections of organic material (particularly vitrinite) are examined to determine reflectivity which is a function of maturity. Pressure - temperature conditions can then be determined and hence the type of petroleum likely to be found. Any organic material is also analysed to determine the amounts of various carbon chain length hydrocarbons which also indicate what type of petroleum is (or was) present. For further information on this topic, see Chapter 3.

Well site analyses include gas detectors which determine the type of gas recovered.

Stratigraphic Investigations

Recent research suggests that the distribution of certain trace elements (particularly carbonate carbon and organic carbon and manganese) may be valuable in clarifying stratigraphic problems. Variations of carbonate carbon in "non-carbonate" rocks such as non-marine shales are often apparently independent of lithology. Such variations may be climatically controlled and thus may allow the construction of time lines in cross-sections of sedimentary basins.

GEOPHYSICAL LOGGING

Introduction

Geophysical logging is usually carried out by lowering a transmitter-detector, called a "sonde", down the borehole. Measurements are usually made during the return journey, the information reaching the surface recorders through a multi-cored cable. Accurate correlation with depth is possible.

Some geophysical logging is also carried out in the laboratory determining such properties as density and magnetic susceptibility.

A mechanical measurement of uncased holes (the caliper log) is valuable in determining intervals which are:

1. Poorly consolidated and "cave"

2. Permeable and which therefore have a cake of filtered drilling mud built up in the borehole

Electric Logs

Measurement of apparent resistivity is valuable in both petroleum and mineral exploration. In the former case, electric logs can be used to determine the "formation factor" which is expressed as:

$$F = \frac{\rho}{\rho w} \quad \ldots\ldots\ldots\ldots\ldots\ldots(13.1)$$

where F = Formation factor
ρ = resistivity of the rock when fully saturated
and ρw = resistivity of the water

Once the formation factor is known, the porosity can be determined from the relationship:

$$F = \frac{a}{\phi m} \quad \ldots\ldots\ldots\ldots\ldots\ldots(13.2)$$

where ϕ = porosity
and a and m are constants depending on the nature of the formation. As m is about 2, the formation factor is approximately inversely proportional to the porosity.

The main disadvantage of electric logs is that they must be run before the hole is cased. There are a number of different methods of measuring electrical properties down a borehole. These are:

1. "Lateral" log, where the potential electrodes are close together and between the current electrodes
2. "Normal" log, where one potential electrode is earthed at the surface and the other potential electrode is between the current electrodes. With short current electrode spacing thin beds are easily detected; with longer spacing, the apparent resistivity (unaffected by the borehole) of the rocks is measured
3. "Micro" log, where electrode spacings are of the order of only a few centimetres which enables determination of the true formation resistivity
4. "Laterolog" measures the resistivity of a sheet-like volume which results in penetration without loss of resolution

5. "Spontaneous Potential" (SP) log measures the potential difference between a single electrode and a reference electrode at the surface. This method is useful in determining "shaliness"
6. "Induction" log is an electromagnetic technique which gives good penetration and allows easy estimation of true formation resistivities

Radiometric Logs

Gamma logs record the natural γ-radioactivity of the rocks through which the borehole passes. Shales (with K40 in clays) and potash-rich evaporites have higher y-counts than sandstones or limestones. The radioactivity is expressed as American Petroleum Institute (API) arbitrary units. Clean sands have a count of about 20 to 40 units and shales have a count of 100 to 140 units depending on area and author (for example, Thornton, 1979). Coals have a similar count to that of sandstones, that is, about 40 units.

Gamma-gamma, or density logs, measure the amount of transmitted radiation which is reflected back by the formation. The amount reflected back is proportional to the density of the formation. If an assumption is made about the average density of the rock-forming minerals of the formation then an estimate of porosity is possible.

Neutron logs, of which there are a number of varieties, estimate the hydrogen content of the formation. Neutrons are passed into the formation from a source at one end of a sonde. Collision with the hydrogen nucleii in the formation reduces the energy of the neutrons and it is this energy which is measured at the other end of the sonde. Hydrogen nucleii occur in water and hydrocarbons in pore spaces so that an estimation of porosity is possible. Naturally, coals also give a good response.

Sonic Log

The sonic, or continuous velocity, log records the time taken by an ultrasonic pulse to traverse a given distance in a formation. This, the interval transit time, is the reciprocal of the velocity of the pulse and is recorded as microseconds per foot (sic). It is a function of many features of the formation including matrix composition, porosity, pore fluids, and degree of compaction and cementation (Allen, 1975). As a generalisation, sandstones and shales have an interval transit time of about

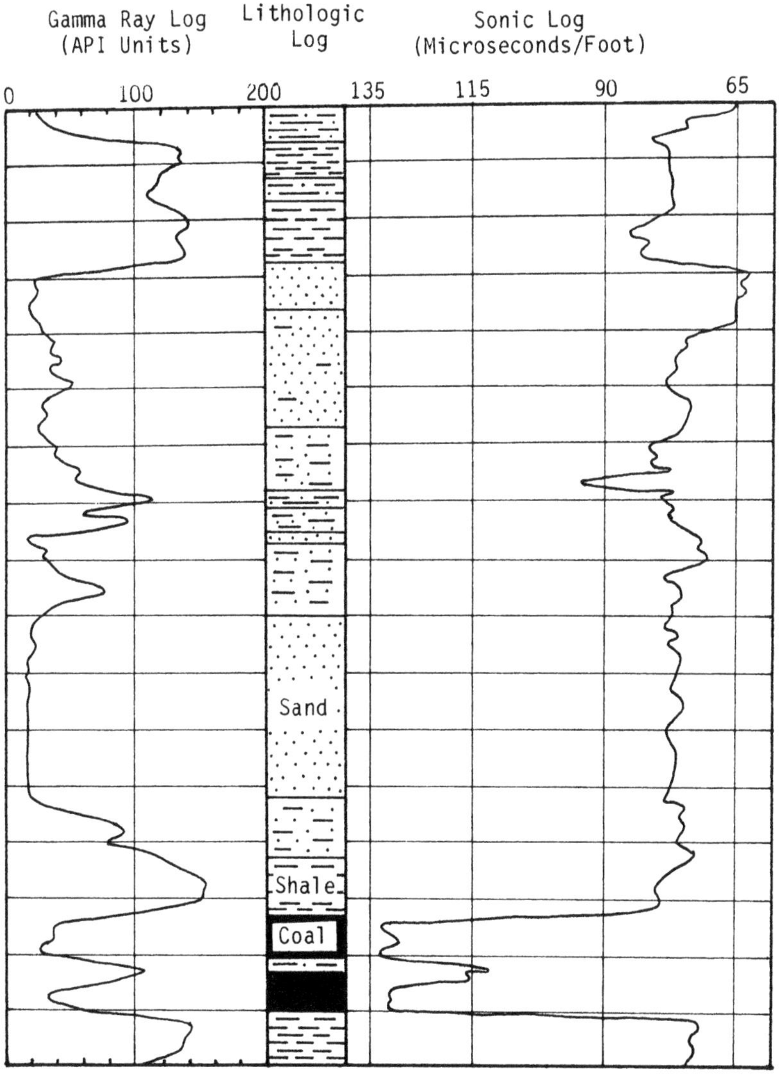

Figure 9.1: The Relationship Between Gamma Ray and Sonic Logs and Lithology.

80 microseconds/ft., whilst the interval transit time in coal is at least 100 microseconds/ft. (Thornton, 1979). Figure 9.1 shows the relationship

between gamma ray and sonic logs in various lithologies.

Magnetic Log
Down hole measurements of magnetic susceptibility are carried out during exploration for iron ore. However, laboratory measurements of core are still more widely used and give more detailed information.

Laboratory Measurements of Physical Properties
Once core is in a laboratory, detailed and accurate determinations of density, magnetic susceptibility, porosity and permeability can be made.

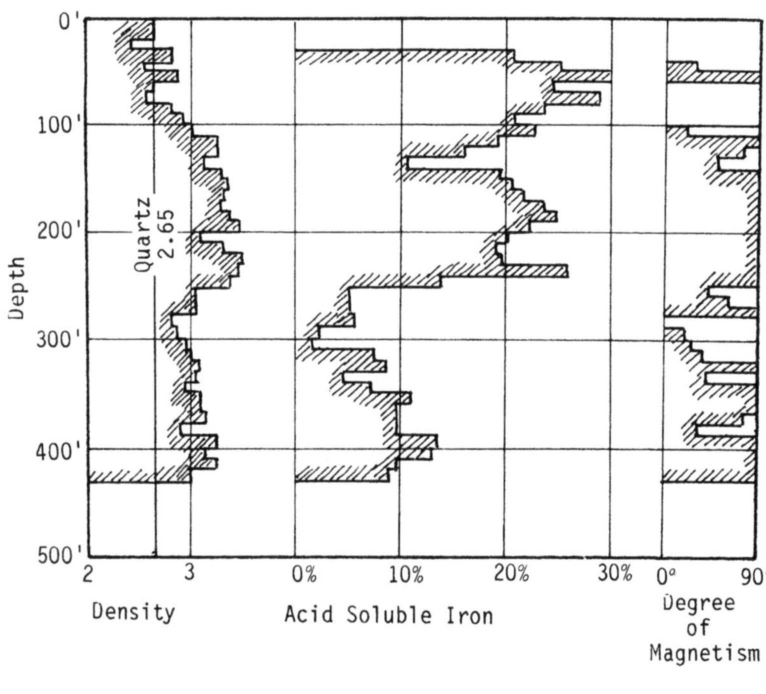

Figure 9.2: The Relationship Between Density, Acid Soluble Iron and "Degree of Magnetism". (After Shackleton, 1966).

A simple estimation of the "degree of magnetism" can be made by suspending a horse-shoe alnico magnet on a 250mm string and determining the maximum angle a piece of core can "lift" the magnet away from the vertical. Figure 9.2 shows the relationship between the "degree of magnetism" (as determined in the field), density and acid soluble iron in diamond drill core in Banded Iron Formation.

QUANTITATIVE LOGS

Data obtained by the various logging techniques can often be used to estimate the quantity of the various elements in the rock passed through by the hole. Magnetic, gravity and resistivity data have long been used in this manner. More recent developments in various types of radiometric logging have increased the range of elements which may be estimated quantitatively.

MECHANICAL LOGGING

Various techniques are applied to consolidated and unconsolidated material to determine parameters which are of engineering value. One of the simplest is the penetration rate referred to in Chapter 12.

Unconsolidated material can provide data such as size-frequency distribution of clastic material, plastic and liquid limits, shrinkage factor and compressive strengths. Such tests are usually carried out on core.

REFERENCES

Allen, D. R., 1975. Identification of Sediments - Their Depositional Environment and Degree of Compaction - From Well Logs, in Chilingarian, G. V. and Wolf, K. H. (Eds.), Compaction of Coarse-Grained Sediments, 1. Elsevier, Amsterdam.

Allman, M. and Lawrence, D. F., 1972. Peels and Embedding, in Geological Laboratory Techniques. Blandford Press, London.

Fitch, A. A. (Ed.), 1982. Developments in Geophysical Exploration Methods - 3. Applied Science Publishers, London.

Glenn, W. E. and Hohmann, G. W., 1981. Well Logging and Borehole Geophysics in Mineral Exploration. Economic Geology, 75th Anniversary Volume, 1981: 850-862.

Bore Hole Logging

Griffiths, D. H. and King, R. F., 1981. Applied Geophysics for Geologists and Engineers (2nd Ed.). Pergamon Press, Oxford.

Parasnis, D. S., 1979. Principles of Applied Geophysics (3rd Ed). Chapman and Hall, London.

Morris, R. C. and Ewers, W. E., 1978. A Simple Streak-Print Technique for Mapping Mineral Distributions in Ores and Other Rocks. Econ. Geol., **73**(4): 562-566.

Shackleton, W. G., 1966. Some Iron Ore Deposits Of the Southern Lincoln Uplands, South Australia. Unpub. Thesis, Univ. Sydney.

Thornton, R. C. N., 1979. Regional Stratigraphic Analysis of the Gidgealpa Group, Southern Cooper Basin, Australia. Bulletin 49, Geol. Surv. S. Aust., Adelaide.

FURTHER READING

JOURNALS

Mining Annual Review
Oil and Gas Journal

Chapter Ten

EXTRACTION TECHNIQUES

INTRODUCTION

Early mining operations include the turquoise mines of the Sinai Peninsula and the flint mines of Norfolk, England. In the Sudan, probably over 4000 years ago, some of the early placer mines were over 250 square kilometres in areal extent and were worked to a depth of two metres (Lewis and Clark, 1964). The Phoenician, Greek and Roman Empires were largely dependent on mines for their continuation. The scale of the ancient tin and copper mines of Cornwall and the alluvial gold and base metal deposits of Spain are testimony to the need and scale of ancient mining operations. The Rio Tinto mine in the Huelva district of Spain appears to be the oldest mine in the world still in production.

The fall of the Roman Empire resulted in a decline in mining activity until the Industrial Revolution in the middle of the nineteenth century, and the discovery of gold in North America and Australia in the middle to late nineteenth century. The Industrial Revolution resulted in the first major shift from the use of man and animal power to mechanisation. The two main inventions leading to this development were the steam engine and dynamite. Although solution mining has been used for thousands of years the development of this and the leaching technique are the most recent and significant advances in mining methods. Recognition of the role of bacteria in these processes has led to the concept of microbiological mining. Table 10.1 is a simple classification of extraction techniques.

The main principle of mining is to extract as much of the resource bearing material and as little as possible of the barren material for as little cost as possible. This principle has not changed

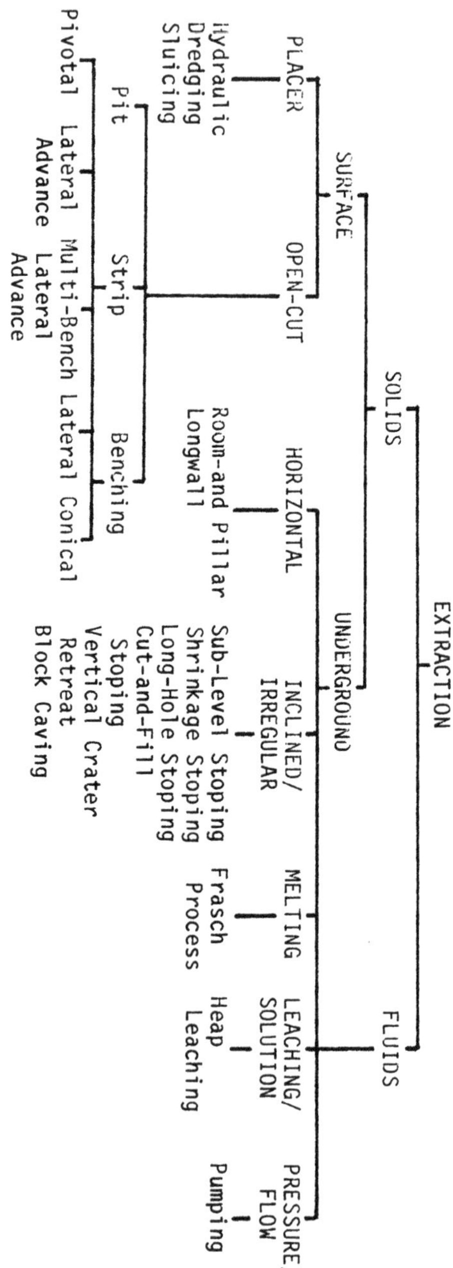

Table 10.1: A Simple Classification of Extraction Techniques.

through the history of mining although initially the main costs were in terms of human lives whereas now the costs are in terms of machinery and power. In either case, efficient mining operations make good use of gravity.

Lewis and Clark (1964), Cummins and Given (1973) and Thomas (1983) are three comprehensive texts dealing primarily with solid material mining methods. Brearly and Atkinson (1968) is an excellent paper on opencast mining techniques. Peters (1978) and Reedman (1979) both have chapters on mining techniques and ore reserve calculations. Several chapters in Giuliano (1981) cover petroleum extraction. Details of recent developments are in the proceedings of the many Institutes of Mining and Petroleum Engineers and Geologists.

SELECTION OF EXTRACTION TECHNIQUE

In the past there were clear-cut distinctions between the extraction of solid rock material by the conventional mining techniques and extraction of resources such as petroleum and water by drill holes. However, as more sophisticated techniques are developed and the demand for minerals increases, combinations of the above techniques are being increasingly used. For example, there are several proposals for recovering oil by mining techniques rather than the normal fluid extraction methods (Dick and Wimpfen, 1980). In all cases, an accurate, detailed knowledge of the geology of the deposit is essential before selection of the most economic technique.

In particular, the choice between surface mining or underground mining is often very difficult. As a generalisation, a horizontal, stratified mineral deposit, such as coal, if covered by overburden the thickness of which is no more than ten times the thickness of the coal seam, would be mined by one of the appropriate surface mining methods. Irregular metallic deposits may be mined by an open cut method if near the surface, but by an underground method if deeper in the crust. It is therefore the form, attitude and depth of a mineral deposit which has the greatest effect on the choice of the suitable mining method (Figure 10.1).

The historic trend is towards mining larger tonnages of lower grade ore. This has been made possible by the rapid increase in mechanisation, particularly in the use of trackless machinery (Morovelli and Karhnak, 1982). Consequently, new mines tend to be large opencast operations which give economies of scale.

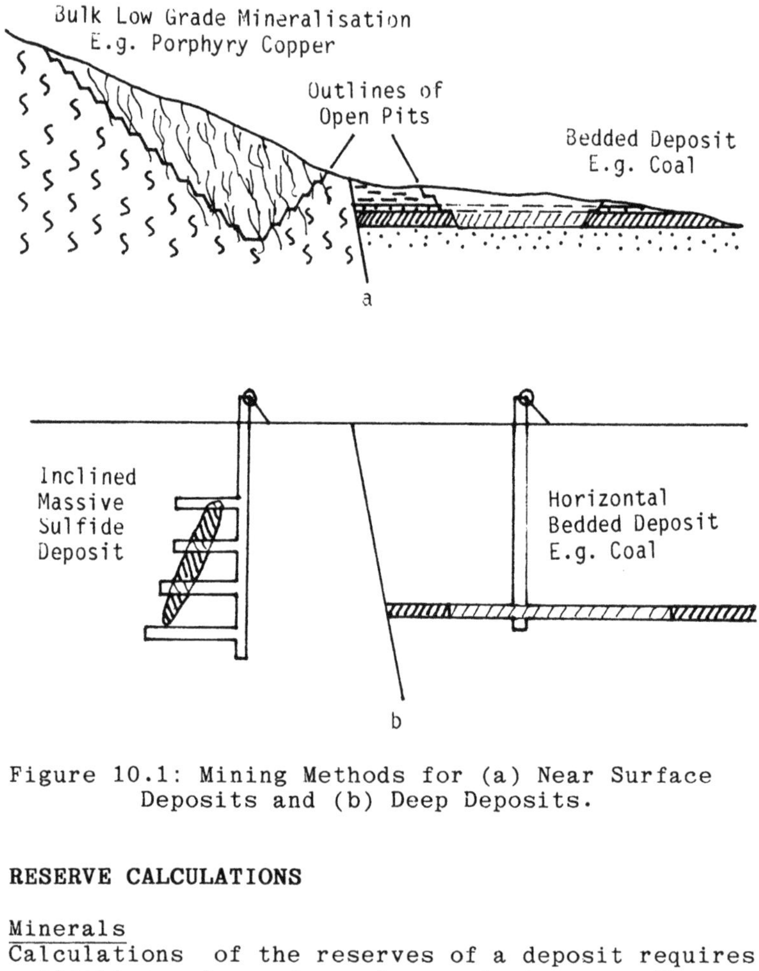

Figure 10.1: Mining Methods for (a) Near Surface Deposits and (b) Deep Deposits.

RESERVE CALCULATIONS

Minerals

Calculations of the reserves of a deposit requires estimations of grade, size and tonnage. The raw analytical data is provided by analysis of drill hole cuttings or core or chip samples of the exposed ore. Linear measurements are provided by normal surveying techniques and tonnage estimates depend on ore density determinations.

Simple arithmetic averaging of analytical data usually provides very misleading information. An intersection of a mineralised zone which has variable grade thicknesses is illustrated by Figure

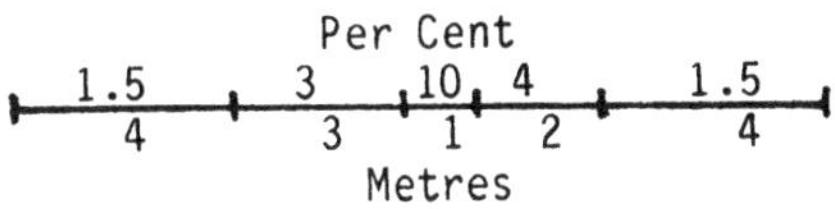

Figure 10.2: Assay Values Across Mineralisation.

10.2. A simple average grade g of the mineralised zone is:

$$g = \frac{\Sigma P}{N} \qquad \text{.................(10.1)}$$

where P = the grade of an intersection and
N = the number of intersections.

In this case g = 5.00%.

However, this method does not take into account the fact that there is very little of the high grade material and there is proportionately more of the lower grade material. It is therefore important to "weight" the average for the length of intersection of each assay value. This is done by:

$$g = \frac{\Sigma(PL)}{\Sigma L} \qquad \text{.................(10.2)}$$

where L = the length of the intersections.

In this case g = 2.8%. This illustrates the large differences in grade that can arise when calculated by the simple rather than the weighted method.

A simple method of determining the weighted average of the shaded area of mineralisation shown in Figure 10.3 is:

$$g = \frac{\Sigma(PT)}{\Sigma T} \qquad \text{..................(10.3)}$$

where P is the grade of intersection T.

Determination of the reserves of a solid body is more complicated. Ore reserve calculations with worked examples are given in Reedman (1979). All these methods are really approximations of reality. The only time that a mining engineer "knows" the real grade of the ore body is when the body has been mined out. The cut-off grade is the lowest grade of ore that can be mined economically. If lower grade material is mined then it is said to "dilute" the ore. Determination of reserves in terms of tonnage requires an accurate knowledge of the density of the various grades of ore. Obviously, higher grade ore will have a higher density and less volume will be required to produce a similar tonnage of mineral than lower grade ore.

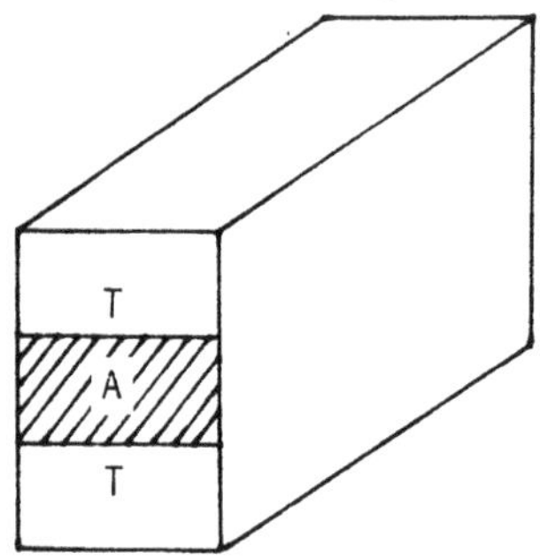

Figure 10.3: Hypothetical Block of Ore.

Fluids
Determination of fluid reserves, particularly petroleum, is dependent on a knowledge of the porosity, structure and other drill hole data of the reservoir. In a simple domal structural trap, at least three drill hole intersections are required to determine the positions of the gas/oil and oil/water interfaces. If the porosity is known the volume of petroleum in place can be calculated. This is not the recoverable petroleum because of factors such as low permeability and high viscosity.

SURFACE MINING

Placer mining
Hydraulicking. Large, unconsolidated heavy mineral deposits, such as gold bearing gravels, may be worked by loosening the gravels with high pressure water jets. A plentiful supply of water at high pressure is necessary. This requires very heavy pipes and a heavy duty nozzle called a "giant", both of which must be made secure. The loosened gravel is transported by the return water to the sluicing operation. This method of mining was widely used in the early gold rush areas of North America.

Dredging. There are two main types of dredges, mechanical and suction. The early dredges were of the mechanical type and were also mainly used in working placer gold and tin deposits throughout the world. Although dredges were rarely used in lakes and rivers, the majority were, and are, operating in their own artificial lakes which they move around with them (Figure 10.4). This applied to the

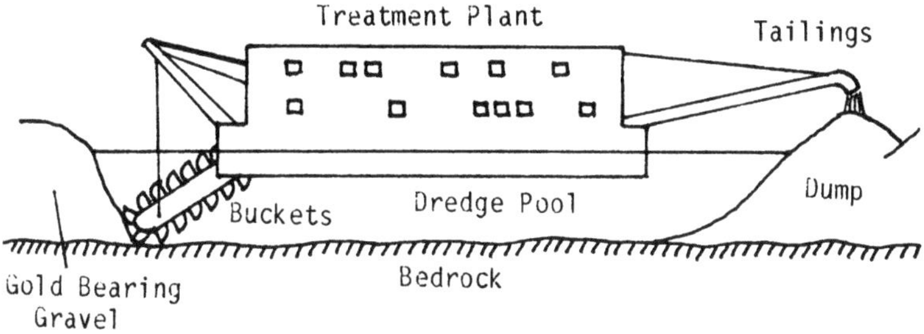

Figure 10.4: Mechanical Dredge.

dredging operations in alluvial gold bearing gravels of North America and Australasia, and still applies to the beach sand mining operations on the east coast of Australia. This type of mining causes considerable disruption to the environment both in terms of damage to the land surface and also by polluting river systems by clay. Most dredges also carry with them the treatment plant for recovering the dense minerals such as gold and tin. Mechanical dredges are also used to remove overburden in open pit mining operations.

Dredging offshore is becoming of great importance. The United Kingdom now obtains a large proportion of her sand and gravel needs from dredging operations around her coast. Similar operations are recovering diamonds off the west coast of Africa. More sophisticated dredging techniques are being developed to recover the vast reserves of manganese nodules lying on the deep ocean floor.

Sluicing. Sluicing itself is strictly not a mining operation but part of mineral technology. This is because the sluices are used to recover the sought after minerals and do not play any significant part in the actual extraction of the ore bearing material. In its simplest application, one or more miners move the ore bearing material by shovel and barrow to the sluice boxes where the material is washed by flowing water. Riffles, or bars at right angles to the flow in the bottom of the sluices, trap the more dense minerals as they are transported along the bottom. The supply of the gravels to the sluices is carried out by hand, front-end loaders, hydraulicking or dredging.

Open-cut Mining

Pit Mining. This is the simplest type of open-cut (or opencast, or open-pit) mining. It is used to mine unconsolidated clays, sands and gravels for engineering and construction purposes. All that is required, on a small scale, is a front-end loader and trucks to remove the material (Figure 10.5). In the case of partly consolidated materials, such as bauxite, some ripping, or tearing up of the ground by bulldozers may be necessary prior to extraction.

Figure 10.5: Loading Gravel from a Pit.

Strip Mining. If overburden is not too thick, strip mining is most commonly used. The principle is to remove barren overburden to expose the essentially horizontal resource for extraction. The majority of strip mines are in "soft rocks", such as shales and sandstones, and the main product is coal. The overburden may need drilling and blasting before removal. In the simplest case, a walking dragline removes the overburden and "casts" the material onto a worked out part of the mine. The exposed resource may also require blasting before removal by a power shovel. A very good example of this type of mining technique is that of the Leigh Creek coal mine in South Australia where the mine advances laterally, or parallel to the working face (Figure 10.6).

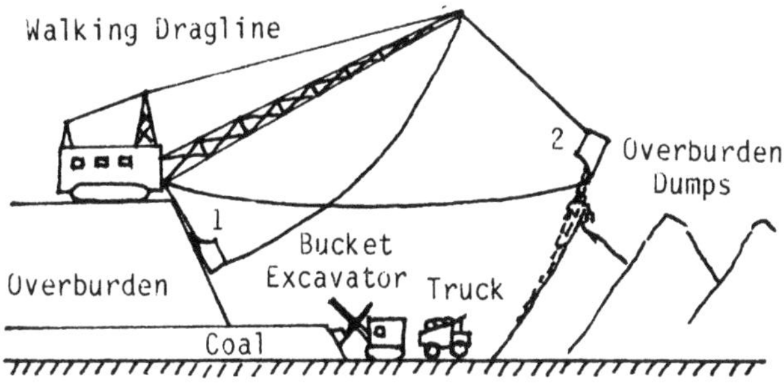

Figure 10.6: Strip Mining by Back Casting.

Rotational, or pivotal, open pits are becoming common especially where it is feasible to set up a permanent treatment and transport plant at the pivotal point. Combinations of pivotal and parallel pits are also being introduced.

Multi-bench, lateral advance mining is similar to the straightforward strip mining method but used where there is greater overburden depth and the necessity for steeper pit walls make it impractical to use draglines. Bucket wheel excavators or dredges are used instead, at least one working on each of the several benches. The overburden is moved to worked out parts of the mine as before, but distances require that the material is transported by truck or, more commonly, conveyor belt systems. The overburden is placed on the dumps by spreaders (Figure 10.7). This system is most

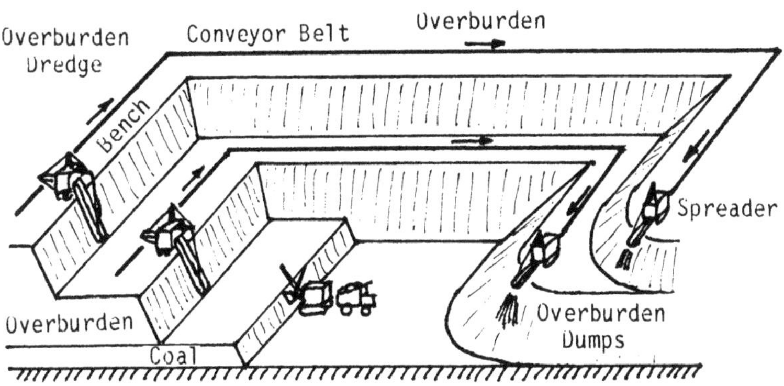

Figure 10.7: Multi-Bench Lateral Advance Mining.

commonly used for coal extraction, for example, the Rhineland, Germany and Yallourn, Australia, brown coal deposits. In the latter two examples the pit is said to advance laterally and the overburden is placed on the mined out area. Both methods may be selected if the overburden ratio (the thickness of the overburden to the thickness of ore) is not greater than about 10 to 1.

Benching. The majority of "hard rock" mineral deposits which occur at or near the Earth's surface are mined by benching. This involves blasting and excavating benches about ten metres high during the removal of overburden and ore extraction. The method is widely used in quarrying operations to produce crushed aggregate and dimension, or building, stone.

Figure 10.8: Bowman's Coal Trial Conical Pit, South Australia.

If the mineral deposit is irregular in shape,

such as porphyry copper deposits, the conical pit method is often used. Conical pits are also used to obtain bulk samples of near surface bedded deposits such as coal (Figure 10.8).

A variation of the strip mining method is used if the mineralisation is moderately to steeply dipping and lenticular. In such cases the benches are parallel to the strike of the mineralisation. This method is commonly used to mine stratiform base metal deposits (Figure 10.9) or iron ore deposits associated with banded iron formations.

Figure 10.9: Mining the Riocin Stratabound Base Metal Deposit, Spain.

If the deposit is small, then the normal open pit methods may be unsuitable. A number of small but very rich base-metal Kuroko type ore deposits in Japan are mined by a novel open pit technique. The deposit is circled by a series of adjacent large diameter drill holes. These are filled with concrete. When the concrete sets, a vertical retaining wall results. The overburden and ore are mined out by a front-end loader and bucket crane (Figure 10.10). If the ore is irregular in plan view, minor underground extraction is possible by cutting through the concrete columns for access.

Figure 10.10: Open Pit Mining of Kuroko Ore, Japan.

UNDERGROUND MINING

Horizontal Bedded Deposits

Room and Pillar. This (and the longwall method) is the characteristic coal mining method. Mining commences from the bottom of the shaft and advances outwards into the deposit. The whole of the seam is mined out with the exception of pillars of coal which are left to support the roof (Figure 10.11). The method results in some 35% of the ore being "tied up" in a supporting role. In many cases, once the primary mining has been completed the pillars themselves are extracted. Extraction starts at the faces furthest from the shaft and moves towards the shaft, the roof being allowed to collapse behind the extraction.

Longwall. Two parallel drives are made into the seam. The coal is mined by a continuous mechanical miner operating between the two drives. Hydraulic supports immediately behind the work face protect the mining operations. As the mining face advances, the roof is allowed to collapse behind into the worked out area. The two drives are retained as access and transport drives (Figure 10.12).

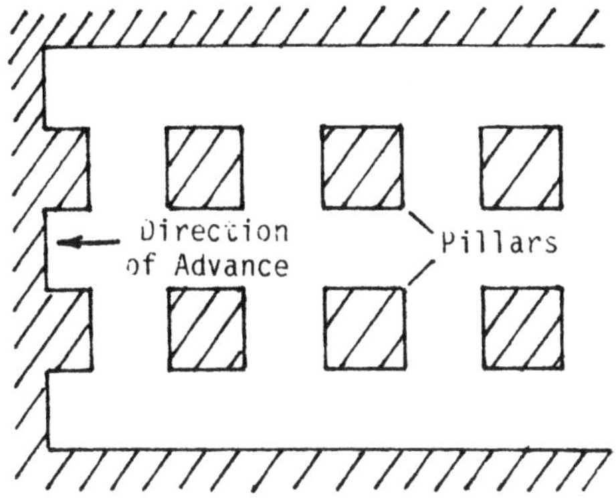

Figure 10.11: Room and Pillar Mining (Plan).

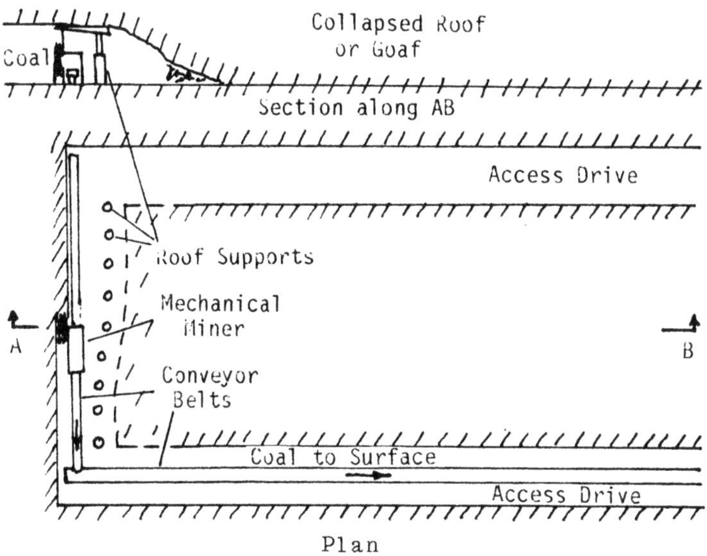

Figure 10.12: Longwall Mining.

Lenticular or Massive Deposits

Introduction There are a large number of methods of extracting such deposits. Some features of underground mining operations are common to many of these methods and these are illustrated in Figure 10.13. Gravity is used as much as possible in

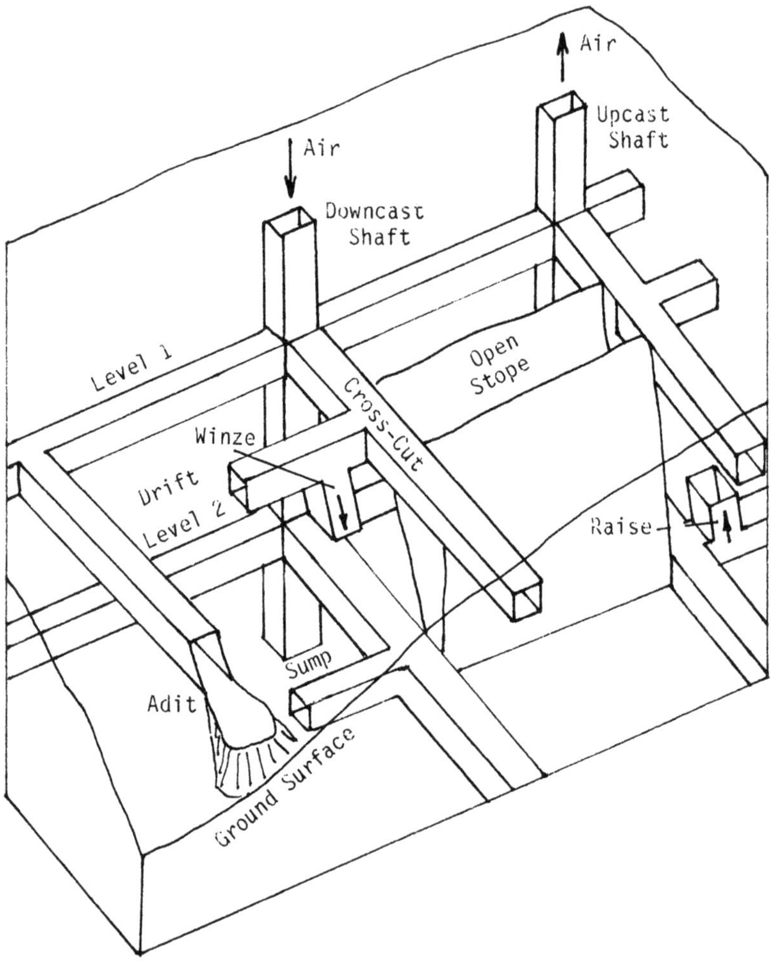

Figure 10.13: Underground Mining Terminology.

mining operations. Many of the vertical openings are now made by machines which essentially are very large drills (Anon., 1983). All the methods described in this section require that the ore must be dipping at a minimum of 50° because the angle of rest of large broken ore is about 45°.

There are many other underground mining methods including block caving, long-hole stoping and vertical crater retreat. Variations on these and other methods are always being developed as are new methods. Continual reference to the literature is necessary to keep up to date. The introduction of trackless machinery has resulted in the increased use of declines, or low angle adit access to many mines. A number of the more common methods are described below.

Sub-Level Stoping. This method is the most widely used. Open stopes are developed from sub-levels. The broken ore is allowed to fall to the bottom of the stope where it is drawn off through the ore passes, or chutes, into the bottom level transport system (Figure 10.14). The grizzlies are large

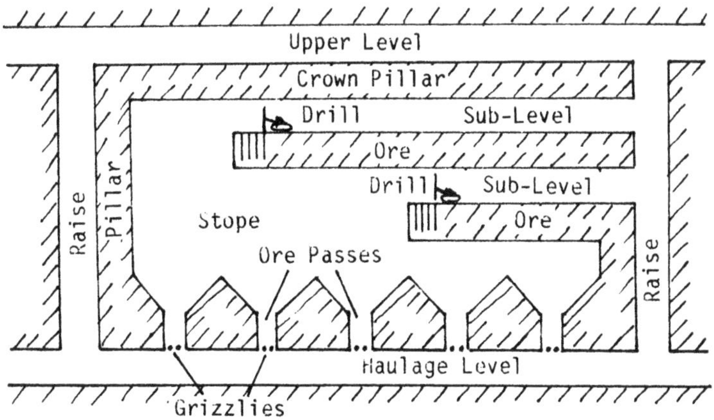

Figure 10.14: Sub-Level Stoping (Section).

screens which retain over-size ore lumps. These lumps must be broken up otherwise the transport system would not be able to cope. This method requires that the wall rocks are strong enough to stand without support when the ore has been removed. Initial extraction is about 60 to 70%, rising to 80% if the pillars are extracted. If all the ore is extracted then dilution with barren rock occurs.

Cut and Fill. The ore is extracted by working upwards from the bottom of the stope. The broken ore falls to the floor of the stope and is scraped into the ore passes. As mining progresses, the ore passes are built up by timbering and the intervening spaces are filled with sand from the tailings of the ore treatment plant (Figure 10.15).

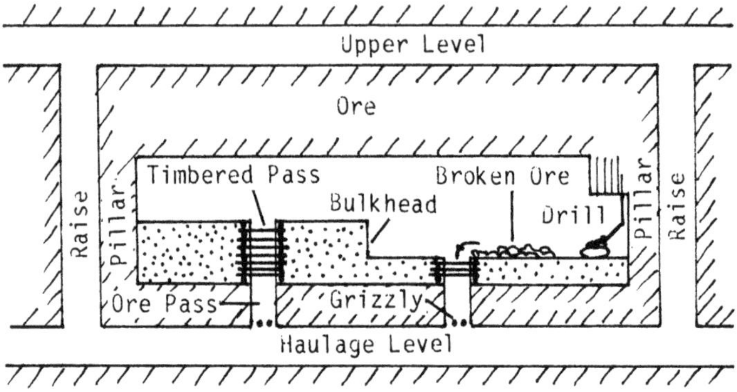

Figure 10.15: Cut and Fill Mining (Section).

This method has the advantage of providing support for the walls of the stope as the ore is removed and also providing a disposal site for the tailings which may otherwise cause an environmental problem.

Shrinkage Stoping. This method is similar to the cut and fill method except that support is given to the walls by the broken ore still in the stope. As the mining progresses upwards, the broken ore is allowed to stay in the stope. Only enough is drawn off to provide a working space for the drilling crews and scraper operators. The floor of the working space is therefore the top of the broken ore. This method has the added advantage of allowing the mine management to cope with variations in production by selectively drawing off ore from this stockpile. It has two disadvantages in that capital is tied up in the broken ore and, in some cases, the broken ore may cement together again and have to be mined again.

FLUID EXTRACTION TECHNIQUES

The extraction of fluids such as oil, gas and water is done mainly by holes drilled into the Earth's crust. The fluid is then either pumped out, or, if under pressure, allowed to flow up and out of the hole.

The case of petroleum raises some interesting engineering and economic problems. In general, the easily located deposits at shallow depths have already been discovered. A great deal of money has to be spent on searching for new fields by means of geological and geophysical methods. When potentially favourable areas have been located, expensive holes have to be drilled, more often than not with little success. High costs are incurred in off-shore drilling where holes are drilled from platforms standing or anchored in water depths of several hundred metres. These cost factors must be taken into account when deciding whether or not to exploit a discovery.

A common method of reducing costs when drilling production wells is to wedge, or deviate, the drill so that several intersections of the producing zone is made. This technique is most often used when drilling from off-shore production platforms (Figure 10.16) and also for drilling off-shore structures from on-shore sites. Small off-shore fields do not justify the expense of

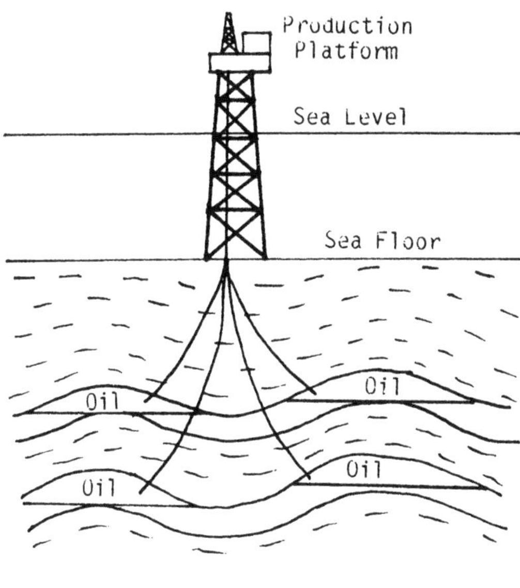

Figure 10.16: Off-Shore Petroleum Production Platform.

production platforms. However, it is often profitable to permanently moor a suitably modified tanker over the field. The tanker is connected to the well head by a riser and stores the produced oil. Ocean going tankers regularly call at the site and on-load the stored oil for shipment to consumers.

The recovery rate for the above primary methods of extraction average about 50%. A number of techniques have been developed to increase the recovery rate. Secondary techniques enhance the ability of the petroleum to flow out of the reservoir by increasing the porosity and permeability or the drive mechanism of the reservoir by essentially mechanical means. These include:

1. Water injection. This method is used when the pressure on the fluid petroleum is not sufficient to raise the petroleum to the surface and it is not considered economic to pump. The petroleum is forced to the surface by pumping water under pressure into the reservoir rocks through specially drilled holes.
2. Fracturing Techniques. The main purpose of this method is to increase the permeability of the reservoir rocks in the vicinity of the oil well. Hydraulic fracturing involves pumping water under high pressure down the well. The high pressure forces apart fractures in the rocks. Spherical quartz grains which are pumped down with the water enter the enlarged fractures and support them when the pumping ceases (Figure 10.17). Ex-

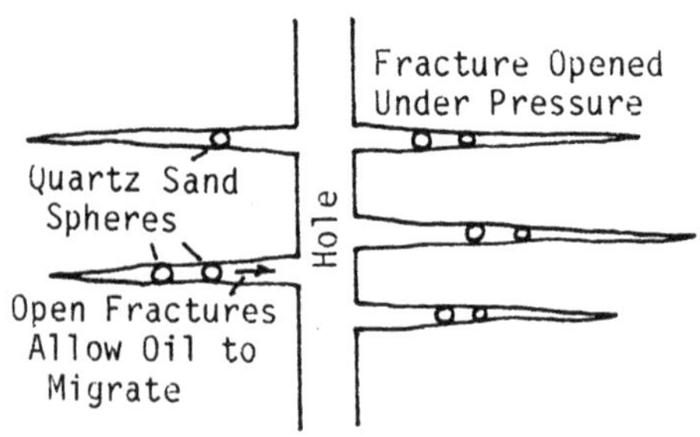

Figure 10.17: Increased Permeability by Fracturing.

plosive fracturing is carried out by detonating explosives down the hole. Acid attack is used where the reservoir rock is a carbonate rock or has carbonate cement.

3. Gas Repressurising. This technique is similar to water injection. Surplus gas from a nearby field is pumped down under pressure into a low pressure pool.

A considerable quantity of oil may still remain in the reservoir. This retention is due to a combination of high viscosity and molecular forces attracting the oil to the mineral grains. Tertiary methods attempt to recover this retained oil by decreasing the viscosity by pumping steam into the reservoir or decreasing the molecular forces by pumping a solvent through the reservoir.

SOLID EXTRACTION BY FLUID TECHNIQUES

Frasch Process

Sulfur has long been extracted by this process. A hole is drilled to the sulfur bed (which is usually associated with a salt dome) and hot water is pumped into the sulfur deposit. The sulfur melts and collects at the bottom of the hole where it is forced to the surface by hot air pumped down the hole under high pressure (Figure 10.18).

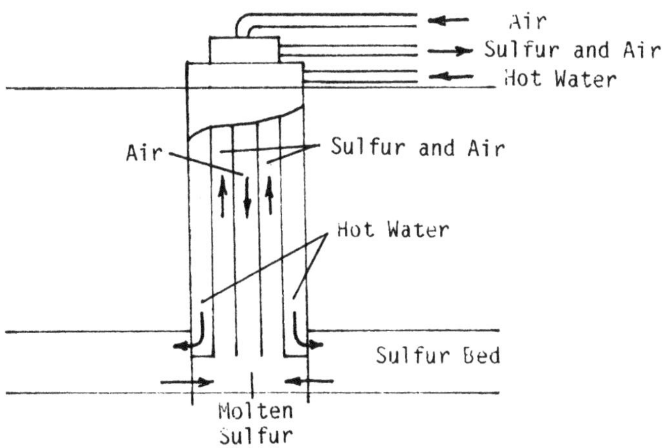

Figure 10.18: The Frasch Process.

In Situ Leaching

A relatively recent development is to leach the required material by using suitable solvents and pumping the resultant solution to the surface where the required material is extracted. This method has been suggested for several bedded uranium deposits in Australia. Provided that permeability and porosity are suitable this extraction method has several advantages over conventional mining techniques. For example, low grade deposits may be worked economically because of the lower costs. Also, there is less environmental impact because no open cut mining operation is needed and spent liquid can be pumped back into the deposit.

A similar technique is to pump metal-rich mine waters to the surface and pass them over scrap iron. This method is used mainly in copper recovery. The copper precipitates and the iron passes into solution (Figure 10.19).

Figure 10.19: Copper Recovery from Mine Waters, New South Wales.
Note the concrete leaching tanks in the foreground and the recovered copper sludge in the metal containers in the background.

In the cases of deposits with low porosity and permeability, such as porphyry copper deposits, experiments have been carried out in the USA by exploding nuclear devices in the deposit. However, the solutions containing copper were considerably radioactive, and such a mining method may require some considerable time lag between fracturing and exploitation.

<u>Heap Leaching</u>. A variation of the in situ leaching method is to recover the metals from crushed

ore stored at the Earth's surface by percolating acidic water through the heaps. Recent work suggests that microbiological activity is of great importance in making this technique highly effective (Brierly, 1982). The metal dissolves in the water and is subsequently recovered by processes similar to the copper/iron interchange. A new development is the carbon-in-pulp technique for gold recovery which, strictly speaking, is a branch of mineral processing and so will be considered in Chapter 11.

ENVIRONMENTAL IMPACT OF MINING

It is fashionable nowadays to criticise the mining industry for the effects of its operations on the Earth's surface. Certainly, in the past the industry was guilty of gross abuse of the environment (Figure 10.20). Today, a keener

Figure 10.20: "Moonscape" in Cornwall. The Dumps from the Treatment of China Clay Workings.

appreciation of the problems arising from mining activities has led to corrective measures being taken both during and after mining. During most modern mining operations, the physical appearance of the mine is enhanced by modern building design and by screening the operation from sight by building embankments or by planting screens of trees. Noise and dust are also suppressed.

After the mine has closed, the worked area can be returned to its original condition or landscaped to an acceptable alternative condition. In the latter case, the area may become an asset of a different kind. Many quarries are used for recreational activities such as boating. The mine may be kept as a living museum and contribute to the education of the nation (Figure 10.21). New mining methods as described above will also reduce the disruption to the environment.

Figure 10.21: New Use For Old Mine.

In many parts of the world, legislation ensures that waste dumps and open cut mines are treated so that they can be an asset rather than a liability to the community. The dumps are graded and revegetated - in one case in Western Australia, worked out mineral sand areas produced a profit when rehabilitated because of the success of growing wheat on the area. Old quarries and open cut mines have been variously turned into sports fields and boating lakes. Abandoned underground mines, usually of coal, often collapse in time resulting in subsidence effects at the Earth's surface.

The main problem of waste disposal will always remain with the mining industry and it is up to the industry and the community to ensure that such problems are contained or even turned into assets.

REFERENCES

Anonymous, 1983. Machines for Shaft Sinking and Raising. Mining Magazine, **148**: 283-295.

Brearly, S. C. and Atkinson, T., 1968. Opencast Mining. The Mining Engineer, No. 99: 147-163.

Brierley, C. L., 1982. Microbiological Mining. Scientific American, **247**: 42-51

Cummins, A. B. and Given, I. A. (Eds.), 1973. SME Mining Engineering Handbook (2 vols). Society of Mining Engineers of the American Institute of Mining, Metallurgical, and Petroleum Engineers, Inc., New York.

Dick, R. A. and Wimpfen, S. P., 1980. Oil Mining. Scientific American, **243**: 156-161.

Giuliano, F. A. (Ed.), 1981. Introduction to Oil and Gas Technology (2nd Ed.). International Human Resources Development Corporation, Boston.

Lewis, R. S. and Clark, G. B., 1964. Elements of Mining (3rd Ed.). John Wiley, New York.

Marovelli, R. L. and Karhnak, J. M., 1982. The Mechanization of Mining. Scientific American, **241**: 62-85.

Peters, W. C., 1978. Exploration and Mining Geology. John Wiley, New York.

Reedman, J. H., 1979. Techniques in Mineral Exploration. Applied Science Publishers, Barking.

Thomas, L. J., 1978. An Introduction to Mining (Revised Ed.). Methuen, Sydney.

FURTHER READING

TEXTS

Institution of Mining and Metallurgy, 1983. Surface Mining and Quarrying. Papers Presented at the 2nd International Symposium on Surface Mining and Quarrying, 1983. I.M.M., London.

Lawrence, L. J. (Ed.), 1970. Exploration and Mining Geology. 8th Common. Mining and Metall. Congress, Aust. and N.Z., Aus.I.M.M., Melbourne.

JOURNALS

Mining Annual Review
Mining Journal
Mining Magazine
Proceedings Aus.I.M.M.
Trans. Instn. Min. Metall. (Sect. A: Min. Industry).

Chapter Eleven

MINERAL PROCESSING

INTRODUCTION

Mineral processing (or mineral dressing) is the treatment of natural and artificial mineral matter for the benefit of society. This involves one or more combinations of liberation (or preparation) and concentration processes using different properties to separate unlike materials from each other. The end product is not changed from the original material. For example, the end product after treating a metallic ore deposit is still an ore mineral, not a metal which is produced by the next stage, metallurgy.

Mineral processing is useful because:

1. Lower grades of ore may be mined.
2. Bulk mining rather than selective mining may be used.
3. Unwanted material may be removed near the mining operation thus reducing transport costs.
4. Smelter costs may be reduced.
5. Metal losses in slag may be reduced.

The main disadvantages are, firstly, the various processes cost money and, secondly, some valuable material may be lost during the processing.

Mineral processing is used in treating metalliferous ores, non-metallic ores, artificial and waste products. The same principles are also used in treating foodstuffs and seeds.

This topic is comprehensively covered by Wills (1981) and Pryor (1965). The proceedings of the various International Mineral Processing Congresses (for example, Jones, 1974) provide information on new developments and case histories. Other sources of information are various mining journals.

PREPARATION

Introduction

Preparation begins as soon as the ore is extracted from the mine when the ore undergoes at least one and usually more episodes of size sorting. Size sorting, the process of separating the ore into various size fractions, is usually done by screening or by classifying. Washing the ore may or may not be done in conjunction with size sorting.

The ore is further prepared by liberating the desired mineral from the gangue (or unwanted Earth material) so that subsequent treatment is possible. The most common method is to break up the ore into finer particles by crushing and grinding - the process of comminution. Alternatively, the ore may be in a very fine state and it may be necessary to increase the size by various methods - the process of agglomeration.

Screening

Screening is carried out on wet or dry feed. The screens may be large (such as the grizzlies in mining operations), medium and fine. The different types of screens include trommels (cylindrical), wire and plastic screens and sieves for fine screening. Screening is relatively inefficient in dealing with sizes below 10-30μ because the feed cannot be separated into two distinct fractions.

Classifying

Classifying separates particles by size and density. A commonly used classifier is a cyclone. The feed is introduced into an inverted conical container. A circular motion, or swirling, motion is imparted to the feed by high pressure water or air (hence the term "cyclone"). The larger and denser particles fall to the bottom of the cone and the remaining material is extracted at the top. The technique is based on the terminal velocity of the particles in the cyclone.

Comminution

Crushing Crushing is the preliminary breaking of the ore into manageable sizes. The primary crusher in many mines is often situated underground rather than at the mine head. The more common crushers are:

1. Gyratory crushers where a cone of metal revolves eccentrically within a housing.

These machines are normally used as primary crushers.

2. Cone crushers are modified gyratory crushers and are usually used as secondary crushers.
3. Jaw crushers where one jaw of metal oscillates back and forward in relationship to another thus nipping the falling ore (Figure 11.1).
4. Hammer mills where the ore is cracked by high speed revolving hammers on a spindle.

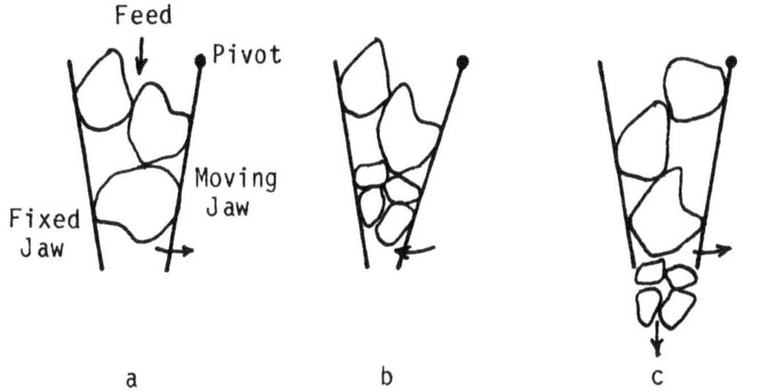

Figure 11.1: The Operation of a Jaw Crusher.
(a) Jaws apart - coarse feed enters crusher.
(b) Jaws together - some ore is crushed.
(c) Jaws apart - crushed material falls through.

<u>Grinding finally</u> Grinding is the process where ore is broken down to a size which is desirable

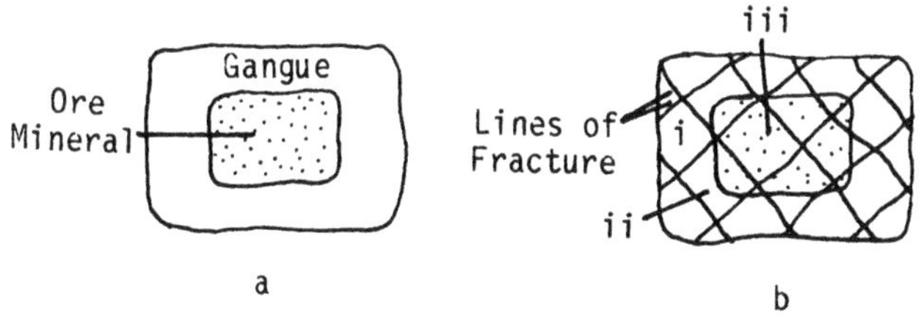

Figure 11.2: Locked Middlings.
(a) Gangue Containing Ore Mineral.
(b) The Result of Crushing:
(i) 100% Gangue.
(ii) Some Gangue, Some Ore Mineral.
(iii) 100% Ore Mineral.

for the separation stage. The liberation of the ore minerals never reaches one hundred per cent. Some locked middlings always remain (Figure 11.2). This problem is not solved by finer grinding because:

1. When mineral grains are less than 10 to 20μ the particles are too fine to be used in most subsequent separation processes.
2. Comminution is a very expensive process. It is estimated that only about 0.1% of the energy used does any useful work.

The more common grinding machines are:

1. Ball mills where the ore is passed through a revolving cylinder containing tough steel balls. These crush the ore by falling on it (Figure 11.3).

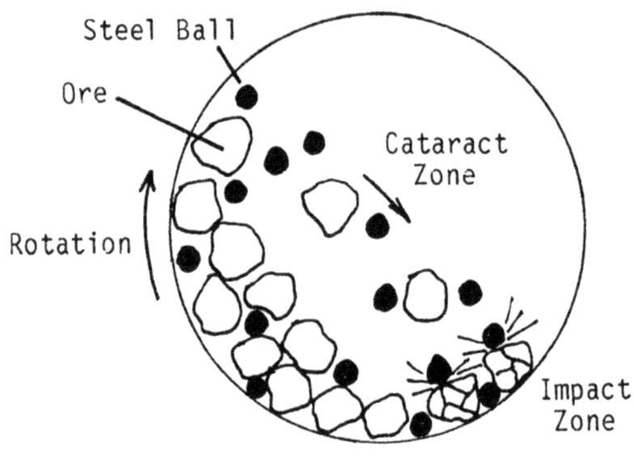

Figure 11.3: The Operation of a Ball Mill.

2. Rod mills are similar to ball mills but crush the ore by nipping the ore between steel rods as they fall on the ore.
3. Autogenous mills where the ore is used to crush itself.

The main problem with comminution (apart from those mentioned above) is that the process cannot produce two distinct fractions. Figure 11.5a is a size frequency diagram of material in two distinct size fractions. Figure 11.5b illustrates what happens in reality. There will always be a variation from that desired. If the required size fraction is x on the diagram, then further grinding will be neccessary to reduce the amount of coarse material (to the right of x). This further grinding will move the whole curve to the left of x and thus

Figure 11.4: The Interior of the Mill, Arinteiro Copper Mine, Northern Spain.
Ball mills of the crushing circuit are in the foreground and froth flotation cells are at the rear of the building.

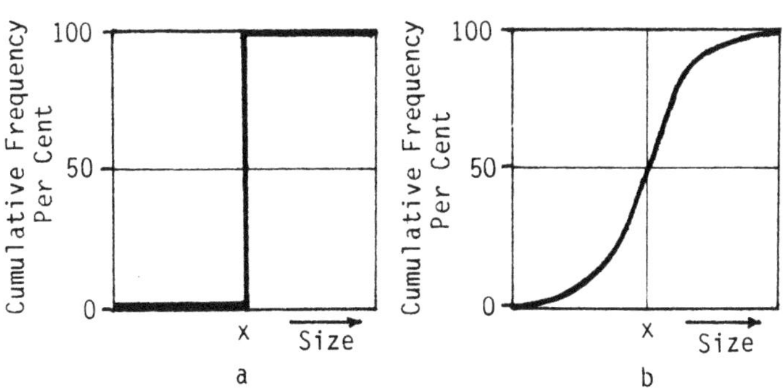

Figure 11.5: Size Frequency Diagrams of (a) Perfect Separation and (b) Reality.

thus larger amounts of finer material will be produced. This illustrates why it is impossible to make a perfect separation in a single step and is the main reason for carrying out sizing at several stages during the comminution process. Oversize material is sent back to the relevant stage to be further reduced in size.

Agglomeration

Agglomeration describes the processes which increase the grain size of the ore. One method is to roast the ore in a process similar to metamorphism.

Alternatively, the fine material can be compressed into particles large enough for subsequent processes. This is called briquetting and is used in the brown coal industry (for example, the Yallourn Valley, Australia). It is also used in the Sherrit Gordon Mine in Canada where nickel concentrate is initially in powder form after leach extraction from the ore.

After initial preparation iron ore is screened to remove the fines which would choke a blast furnace. If the ore produces a high proportion of fines then it may be economic to pelletise the fines to form particles large and strong enough to use in the blast furnace.

SEPARATION, SORTING OR CONCENTRATION

Radiation Sorting

Sorting can be carried out if there is sufficient contrast in the colour of the various materials. It is a simple technique and if labour costs are low or the required material has a very high value (diamonds for example) this method can be carried out by hand.

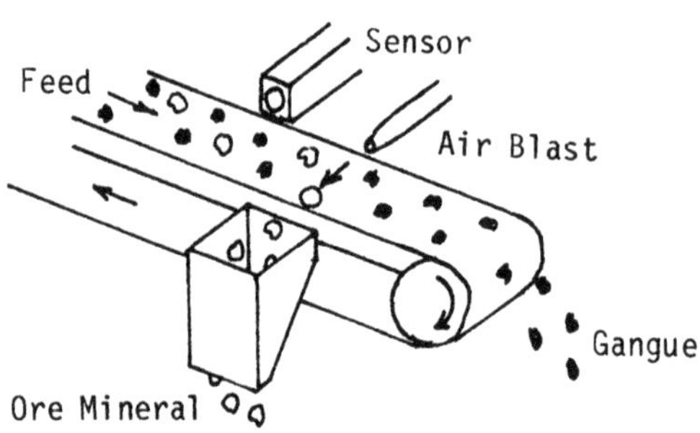

Figure 11.6: Radiation Sorting.

Other radiation sorting uses differences in reflectivity, radioactivity or reaction to X-rays. The ore is passed in front of an appropriate sensor which operates machinery which accepts or rejects ore grade material. This is usually done by a blast of compressed air (Figure 11.6).

Magnetic Sorting

The various magnetic methods depend on the fact that if a mineral is placed in a non-symmetrical magnetic field then the mineral will move. If the mineral is diamagnetic (normally considered to be non-magnetic simply because of the extremely weak magnetic properties) then the mineral will move to the high intensity field. If the mineral is para-magnetic then the mineral will move to the low magnetic field. If the para-magnetic field is parallel to the diamagnetic field then the mineral is ferromagnetic. Most magnetic techniques are based on the differences between para- and ferro-magnetic properties.

Low intensity separators recover highly magnetic minerals. Their most common use is the separation of tramp iron, the rubbish of the mine such as hammers, bolts and so on, which is contained in the run-of-mine ore. If this material enters the crushing circuits much damage may be done. A large electromagnet is suspended over the conveyor belt which delivers the ore to the crusher and removes any magnetic material (Figure 11.7). Another simple

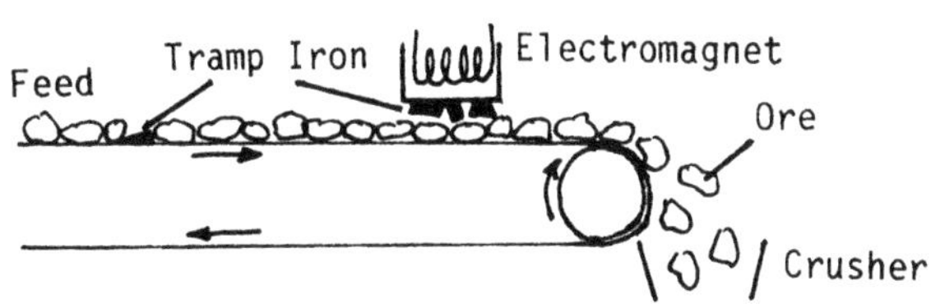

Figure 11.7: Magnetic Removal of Tramp Iron.

dry magnetic separator is illustrated in Figure 11.8. This separates magnetic from non-magnetic ore and/or gangue. It is most effective in separating relatively large grained material - greater than one or two millimetres in diameter. If the material is fine grained, less than 10μ, then low intensity wet magnetic separators are used.

High intensity separators use a high magnetic field with a high gradient to separate "non" magnetic minerals. They are widely used in heavy mineral beach sand plants. In such deposits,

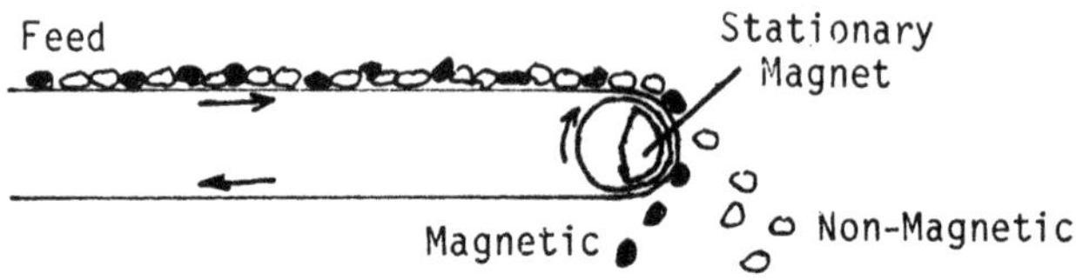

Figure 11.8: Low Magnetic Intensity Separator.

minerals such as ilmentite and rutile are not ferro-magnetic but there is sufficient difference in their magnetic properties to make high intensity magnetic separation possible.

Electrical Sorting

This technique depends on the mineral holding a surface charge. The **electrostatic method** (Figure 11.9) produces the charge by rubbing the mineral

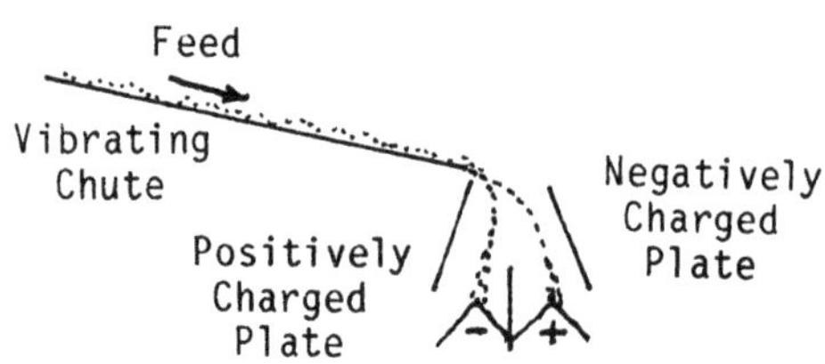

Figure 11.9: Electrostatic Separation.

particles together. This can be done by vibrating the feed chute. The minerals must have fresh surfaces because oxidation greatly reduces the effectiveness of the method.

The **electrodynamic technique** uses differences in the surface conductivity of the minerals (Figure 11.10). An electrical discharge ionises gases in the air. These are sprayed onto the minerals on the drum. At zone A all the minerals are charged. The conductive minerals lose most of their charge to the drum. However, more charge is continually added. In zone B the conductive minerals lose their surface charge and become positively charged. The non-conductive minerals remain negatively charged. As soon as the conductive minerals gain their induced positive charge they will be repulsed by the positively charged drum and they will also be attracted to the negatively charged rod electrode. The non-conductive minerals remain attracted to the drum from which they are removed by gravity or by scraping. This method is used to separate beach

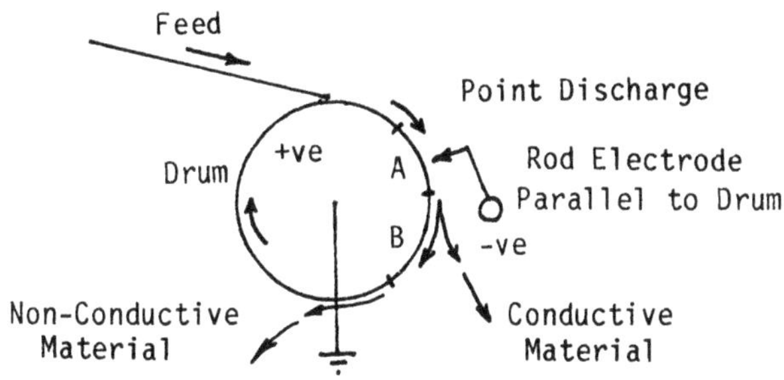

Figure 11.10: Electrodynamic Separation.

sand minerals, coal from slate or shale and hematite from quartz. The feed must be dry and between one millimetre and 50μ in diameter.

Gravity Sorting

Introduction Successful gravity sorting depends on the concentration criterion (c.c.) (Equation 11.1). If the concentration criterion is more than

$$c.c. = \frac{\rho h - \rho f}{\rho l - \rho f} \qquad (11.1)$$

where ρh is the density of the heavy mineral
ρl is the density of the light mineral
and ρf is the density of the fluid medium.

2.5 then separation is possible down to very fine sands. The lower limits of the size of the material which can be treated increases with decreasing concentration criterion until the cut-off value of 1.25 is reached when another separation method must be found.

The gravity method of separation can be carried out in vertical currents using jigs, in flowing films using sluices or by using a combination of these, using shaking tables.

Jigs The minerals become stratified by passing a water current through them. An upward flow is called pulsation and a downward flow is called suction. The process depends on differential initial acceleration and hindered settling. Jigging feed that contains two minerals of different density and a range of sizes densities will result in some segregation (Figure 11.11a).

Figure 11.11: Segregation by Density and Size.
(a) Jig Segregation.
(b) Consolidation Trickling Segregation.
(c) The Net Effect of Combined Jigging and Consolidation Trickling.

If the minerals are trickled down through a bed of coarser material then the denser particles tend to move through the bed more rapidly than the less dense particles. In addition, the smaller particles will move through the bed more rapidly than the larger particles. This is called consolidation trickling and the net effect is illustrated by Figure 11.11b. If consolidation trickling and differential initial acceleration are used together then the two minerals will be separated (Figure 11.11c).

Jigs can provide differential initial acceleration by either pulsing the water with the sieve stationary or pulsing the sieve with the water stationary. The product can be discharged over the screen if the sieve size is less than the mineral grain size, or through the screen where the lower layer of the mineral passes through the screen. The screen must have a sieve size approximately the same size as the mineral grains.

Jigs are most commonly used in the coal, alluvial tin and iron ore industries and are used with relatively coarse feed, about 50mm diameter. If the density of the minerals is low, then different grain size particles must be used. However, if the density difference is large then good separation is obtained no matter what the grain size (as in the case of gold).

<u>Flowing Film Concentration</u> The principle of this method is that if a film of water is travelling down a smooth inclined surface, the velocity is at a maximum at its surface and reduces to zero at the solid surface at the base of the flow. The drag

force down the plane due to the water flow is equal at any one depth on any mineral grain. However, the frictional forces hindering the movement down the plane depend on the mass of the mineral grain. These relationships are illustrated in Figure 11.12.

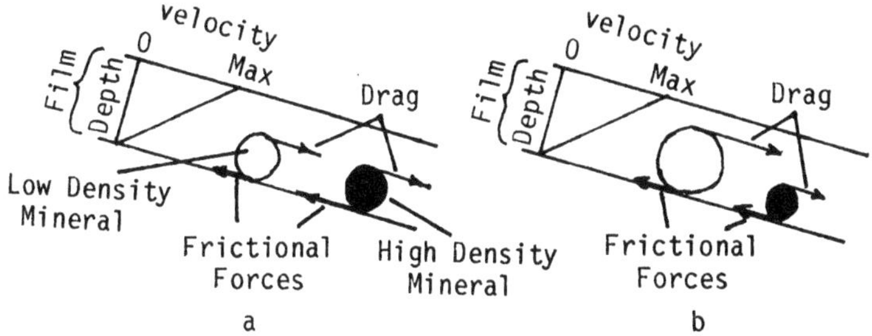

Figure 11.12: The Principles of Flowing Film Concentration.
(a) Grains of Similar Size and Different Density.
(b) Grains of Different Size and Different Density.

If the particles are of different size there is more drag on the larger particles due to the velocity gradient through the film. All the variations of the simple sluice box depends on the flowing film principle (Figure 11.13).

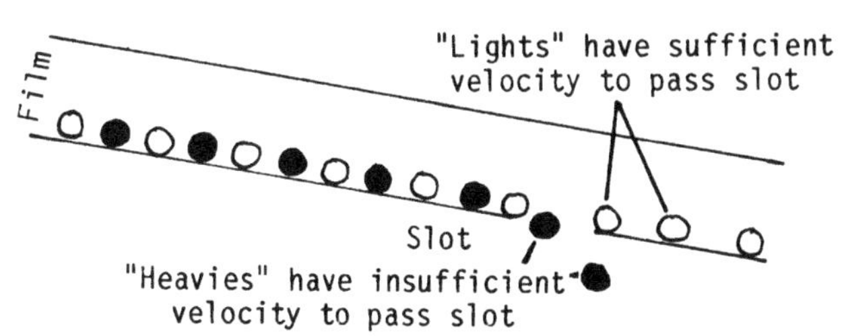

Figure 11.13: Slotted Sluice Box Separation of Heavy Mineral Beach Sand.

A similar separation method is the Humphries Spiral which is used in the concentration of heavy minerals in beach sands. The concentrated beach sand material (after passing through slotted sluice boxes) is washed down a spiral column. Because of the principles outlined above, the light minerals flow on the outside of the curve and the heavies on the inside. The heavy minerals are removed by adjustable arms which direct the mineral grains through a hole into the concentrate bin (Figure 11.14.)

The shaking table is in wide use in the mineral industry and uses the principles of both

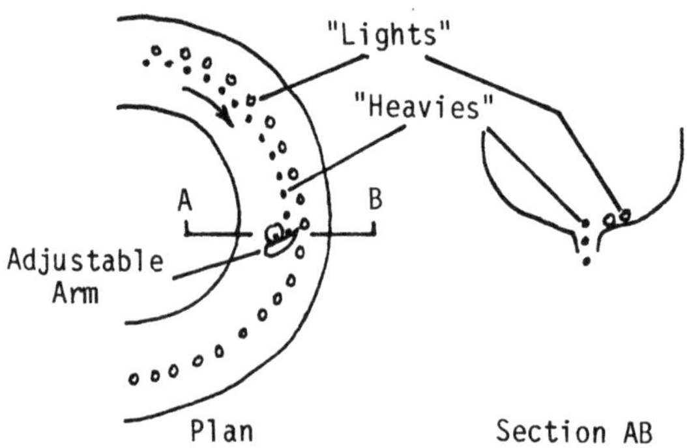

Figure 11.14: The Humphries Spiral.

jigging and flowing film concentration. There are three distinct operations:

1. In the riffle section, the heavy minerals are trapped by the riffles (which are thin slats of wood at right angles to the water flow) as in a normal sluice box.
2. The vertical shaking motion of the table segregates the minerals by the jigging principle. The less dense minerals are washed towards the long side of the table.
3. The sideways shaking motion moves the minerals trapped by the riffles towards the short side of the table. Because the riffles are tapered in this direction, finer mineral grains are also moved to the long edge of the table.
4. In the film sizing area the lighter particles are moved further by the film of water than the heavies.

The result is the separation shown in Figure 11.15.

<u>The Froth Flotation Process</u>

This process was invented to solve the separation problems when the Broken Hill ore body in Australia was first mined. When finely ground mineral mixtures are made into a pulp with water and certain chemicals, some of the minerals will float to the surface of the pulp where they can be removed (Figure 11.16).

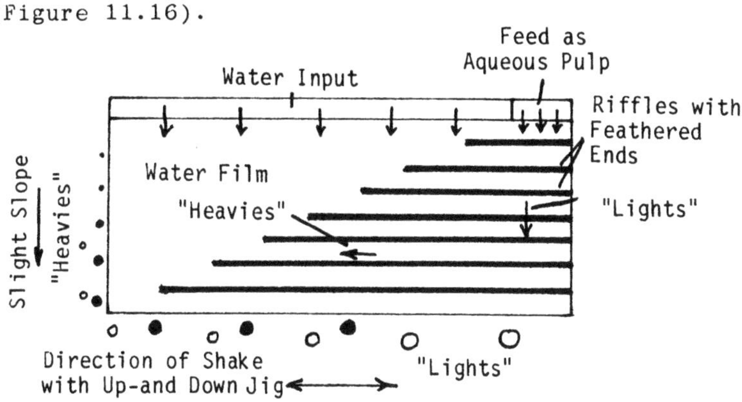

Figure 11.15: Shaking Table Concentration.

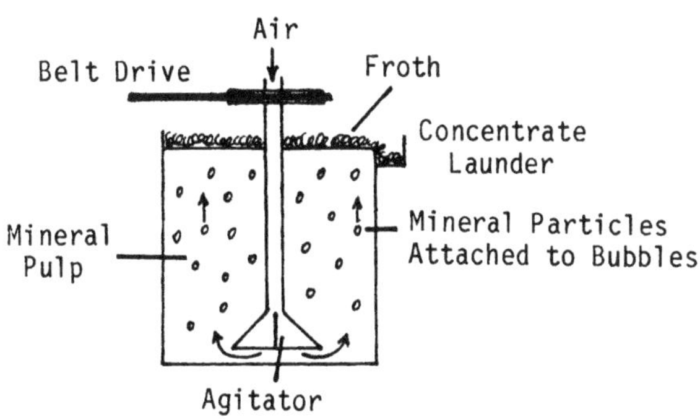

Figure 11.16: A Froth Flotation Cell.

The basis of this method of concentration is the phenomenon of wettability. When certain chemical reagents, called collectors, are added to a mineral pulp, some minerals become less wettable than others. The degree of wettability depends on the surface chemistry of the mineral. Collector reagents are soluble organic heteropolar compounds. One end of the compound is soluble in water because of its polar ionic nature. The other end of the compound is essentially an insoluble non-polar covalent tail (Figure 11.17).

The most widely used collectors are those of

the xanthate family of chemicals. The collectors are concentrated at the mineral/solution interface by adsorption. It is assumed that the active, polar end is attached to the mineral surface and the non-polar insoluble part is in the solution. The

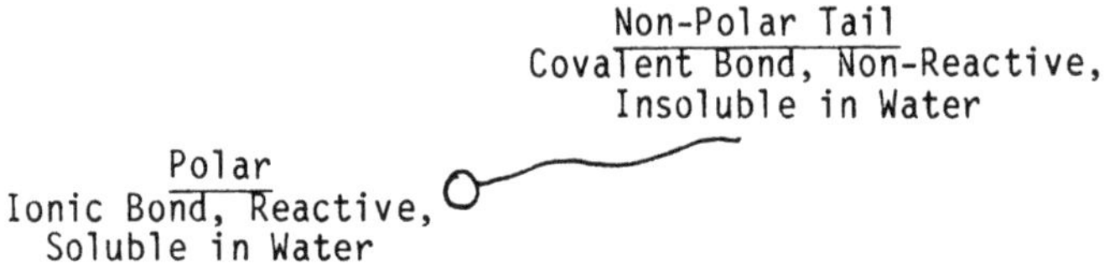

Figure 11.17: The Structure of a Collector.

result is that the mineral surface acts like a hydrocarbon and is thus non-wettable. A measure of wettability is the contact angle (θ) that a drop of solution makes with the mineral surface. The greater the contact angle the less wettable is the mineral (Figure 11.18).

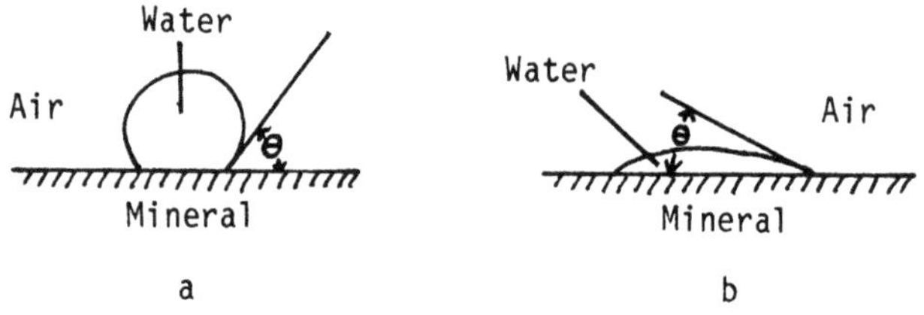

Figure 11.18: The Relationship Between Contact Angle and Wettability.
(a) Low Wettability, High Floatability.
(b) High Wettability, Low Floatability.

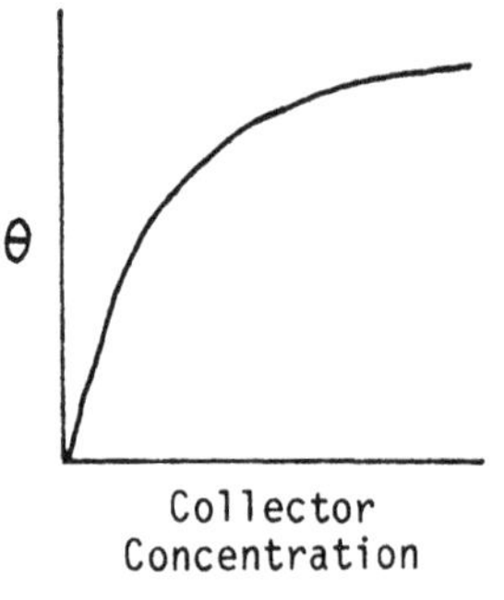

Figure 11.19: Concentration of Collector and Contact Angle.

The contact angle is dependent on the concentration of the collector in solution. The greater the amount of collector the greater the contact angle until a maximum is reached which is different for different minerals (Figure 11.19).

The contact angle is also dependent on the chain length of the non-polar part of the collector. For example, for any one mineral and a collector with the same active portion, the contact angle increases with the chain length until a maximum is again reached (Figure 11.20).

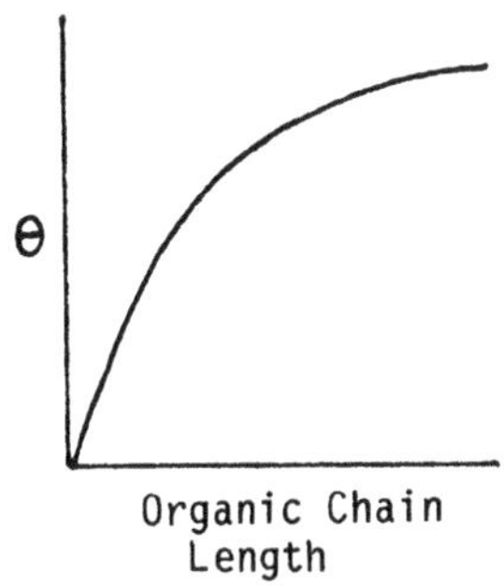

Figure 11.20: Chain Length of Collector and Contact Angle.

Other ions in solution will affect the efficiency of the collector so other reagents are used in conjunction with the collectors. These are:

1. Depressants which act to inhibit the flotation of several minerals at the one time. Depressants act by removal of the collector, competing with the collector, destroying or modifying the collector or removing prior activation.
2. Activators which assist absorption of the collector onto the mineral required to float.

Another chemical added to the pulp is a frother. This is added to stabilise the froth at the surface of the pulp/air interface because of its similar structure to collectors. Frothers normally reduce the surface tension at the interface. However, when the bubble stretches, the frother concentration decreases and the surface tension increases thus stabilising the bubble. The common frothers are alcohols and esters.

The mineral is put into the froth by flotation, splashing and carried by bubbles in a piggy-back fashion. If the bubbles are too tough there is no drainage of the mechanically carried particles and the result is a low grade, high recovery product as in (a) in Figure 11.21. If the bubbles are too weak the mechanically carried particles are drained but there is also excessive drainage of the desired mineral resulting in high grade, low recovery material as in (b) in Figure 11.21.

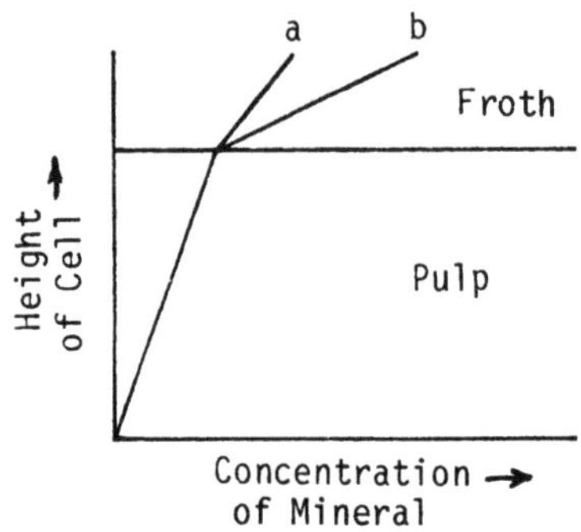

Figure 11.21: Relationship Between Grade and Recovery in a Froth Flotation Cell.
(a) Low Grade, High Recovery.
(b) High Grade, Low Recovery.

To solve this problem, most concentrator plants have what are called rougher and scavenger stages in the froth flotation circuits. There are also cleaner cells to produce a higher grade final concentrate. Figure 11.22 illustrates these units in a simple flow diagram or sheet. The rougher cells essentially operate under the conditions of (a) in Figure 11.21 and the scavenger circuits operate under the (b) conditions.

Dewatering

Nearly all the concentration methods described above involve the use of large amounts of water. This water must be removed from the final product first to reduce transport costs and secondly to recover the water for further use in the plant. This is accomplished by the use of settling tanks or ponds for removal of the water from the tailings and by filters of various types for removal of the water from the concentrates.

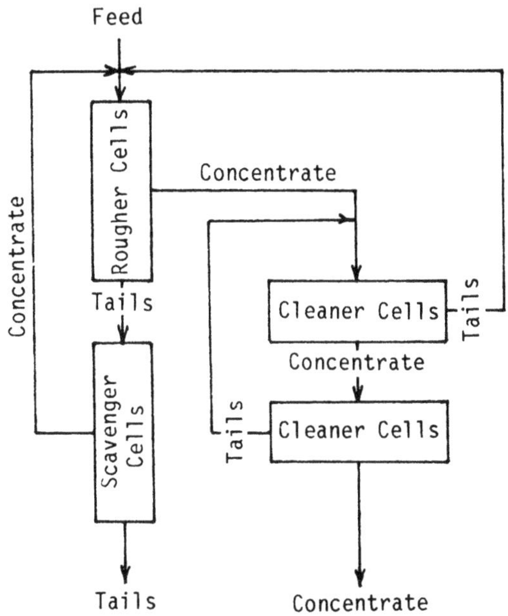

Figure 11.22: Simple Froth Flotation Circuit.

CARBON-IN-PULP

This relatively new technique is included in this chapter because of its general use in gold mining operations. Finely ground gold-bearing ore is made into a thick pulp. This pulp is passed through a leaching tank where a solution of cyanide and caustic soda extracts most of the gold into the solution. The cyanided pulp then passes through a series of carbon-in-leach cells where it encounters a countercurrent flow of activated carbon. Gold is retained by the carbon. The gold is removed from the carbon by another cyanide and caustic soda solution and deposited electrolytically onto steel wool. The steel wool is then dissolved in acid and the gold is filtered, dried and smelted.

DESIGN AND ASSESSMENT OF THE PROCESS

During the feasibility stage of the investigation of a mineral deposit, a bulk sample of the ore should be obtained so that the selection of the mineral separation processes may be relevant to the deposit. There are many instances where insufficient investigations of the minerals and their relationships resulted in poor plant performance.

Ideally, the steps required to produce the most efficient process design are:

1. Identification of all the constituents of the ore. This includes the useful constituents (the products); the useless ones (the tailings); and the harmful ones (those which interfere with the process or spoil the use of the product).
2. Identification of all the mineral properties. These include grain size, texture, hardness, type of cleavage or fracture, and other physical and chemical properties. With this information, the separation criteria, the process can be designed.

Process design includes the following stages:

1. Process selection.
2. Process testing.
3. Process or flow sheet design.
4. Plant design.

Because all processes have their limitations, an accounting procedure should oversee the suit-

Feed
(1000 tonnes, 5%Zn)

Treatment

Concentrate
(100 tonnes, 45%Zn)

Tailing
(900 tonnes, 0.55%Zn)

Figure 11.23: Simple Processing Plant Flow Sheet.

TABLE 11.1

Product	Weight Tonnes	Weight %	Zinc %	Zinc Tonnes	Zinc Distribution
Feed	1000	100	5	50	100
C'trate	100	10	45	45	90
Tailing	900	90	0.55	5	10

Distribution of Materials at Various Positions in Flow Sheet of Figure 11.26.

ability of the process. Figure 11.23 illustrates a simplified flow sheet of a processing plant and Table 11.1 accounts for the material in different parts of the process.

From Table 11.1 the following can be determined:

1. Recovery $= \dfrac{\text{Zn in concentrate} \times 100}{\text{Zn in feed}}$

$$= \frac{450 \times 100}{500}$$

$$= 90\%$$

2. Ratio of Concentration $= \dfrac{\text{Weight of Feed}}{\text{Weight of Concentrate}}$

$$= \frac{1000}{100}$$

$$= 10$$

3. Enrichment Ratio $= \dfrac{\text{Zn\% in Concentrate}}{\text{Zn\% in Feed}}$

$$= \frac{45}{5}$$

$$= 9$$

At least two of the above parameters are needed to determine the suitability of the process. However, ultimately, the only factor that really defines how good the process is, is the profit. Most processes operate over a range of these parameters. For example, the design can opt for higher volume of recovery at lower grade and vice versa. The degree of freedom to produce a profit is relatively small. Computers are used to determine the optimum balance between recovery and grade. The advent of spread sheets makes this a relatively simple exercise.

REFERENCES

Jones, M. J. (Ed.), 1974. Tenth International Mineral Processing Congress (1973). The Institution of Mining and Metallurgy, London.

Pryor, E. J., 1965. Mineral Processing (3rd. Ed.). Elsevier, Amsterdam.

Wills, B. A., 1981. Mineral Processing Technology (2nd Ed.). Pergamon Press, Oxford.

FURTHER READING

JOURNALS

Mining Annual Review
Mining Journal
Mining Magazine
Proceedings Aust. I. M. M.
Trans. Instn. Min. Metall. (Sect. C: Min. Process. and Extract. Metallurgy).

Chapter Twelve

ENGINEERING GEOLOGY

INTRODUCTION

Engineering geology is the application of geological knowledge to solving geological problems encountered during construction of civil engineering structures. It deals with a relatively thin veneer of the Earth's crust - from the surface to a depth of about 100 metres. Geologists concerned with this limited environment must have a suitable knowledge of rock and soil mechanics so that information can be presented in a practical form. This more rigorous quantitative approach to engineering geology and has led to the development of geotechnics.

This chapter primarily reviews foundations, slope stability and dams. Engineering geology and geotechnics are also applied to solving problems in tunnel and canal construction and underground mining operations.

Krynine and Judd (1957), Legget (1967) and Sowers (1979) are three general texts covering the wide range of engineering geology and geotechnics. Capper and Cassie (1960) concentrates on soil mechanics. Peters (1978) has a useful summary of geotechnics.

ENGINEERING PROPERTIES OF ROCKS

Compressive strength

This is the stress required to break a loaded unconfined sample (Figure 12.1). Igneous rocks generally have higher compressive strengths than sedimentary rocks. The highest compressive strengths occur where the compressive stress is normal to bedding and foliation.

Tensile strength
This is the load required to produce cracking in a rock (Figure 12.2). This is generally one tenth or less of the compressive strength and is negligible for soils.

Figure 12.1: Determination of Compressive Strength.

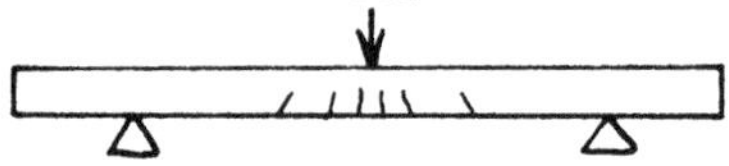

Figure 12.2: Determination of Tensile Strength.

Structure
Faults, joints and bedding planes greatly influence the stability of rocks. In the case of slope stability, the dip of the rocks, together with intersecting joint planes dictates whether or not the slope will fail. In the case of dam foundations, the same structural features will affect the ability of the dam to contain water. Structures within the abutments may be factors in determining whether or not a dam will fail.

ENGINEERING PROPERTIES OF SOILS

Atterberg Limits
The properties of soils (in an engineering sense, "soils" include any unconsolidated Earth material) have long been areas of intense investigation. The engineering properties of any one soil are dependent, to a large extent, on the soil moisture content. The soil passes from possessing one set of properties to another at certain moisture contents called consistency limits (Figure 12.3). Atterberg, a Swedish soil scientist, developed simple tests to determine these limits.

The Atterberg Limits are:

1. The liquid limit which is the minimum moisture content at which the soil will flow under its own weight.
2. The plastic limit which is the minimum moisture content at which the soil can be rolled into a thread three millimetres in diameter without breaking.
3. The shrinkage limit which is the moisture content at which further loss of moisture does not cause a decrease in the volume of the soil.

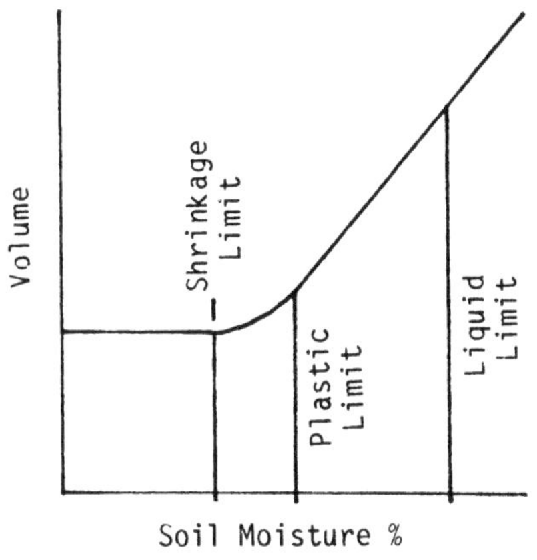

Figure 12.3: Atterberg Consistency Limits.

The plasticity index is the difference between the liquid and plastic limits and is generally inversely proportional to the grain size of the soil. The shrinkage ratio is related to the consistency limits and is simply the rate of decrease per unit volume with decrease in moisture content.

Compressibility

When a load due to a structure is placed on a soil, the soil reacts by decreasing in volume. This is effected by reduction of the pore spaces, or voids, by expulsion of water and air. The vertical downward movement of the structure is called settlement. Settlement is acceptable if it is uniformly distributed underneath the structure. The structure may fail, however, if the amount of settlement varies underneath the structure - a case of differential settlement. A soil which has decreased its volume under compression is said to be consolidated. Compression increases with time until a maximum value is reached (Figure 12.4). If the compression is carried out by rolling or tamping, the soil particles are packed closer together and the soil density is again increased. This is called compaction.

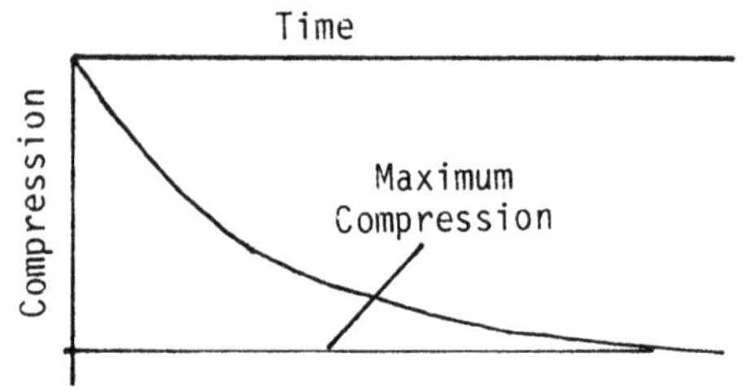

Figure 12.4: The Relationship Between Time and Compression.

After removal of the load the soil may retain its new density (compressive deformation). However, some soils rebound when the load is removed (elastic deformation) or, in the case of clayey soils, there is lateral flow of the soil under load (plastic deformation).

Shear Strength

The maximum resistance of a soil to shearing stresses is called the shear strength. Under conditions of zero confining pressure, cohesionless soils such as sands have low shear strengths and cohesive soils such as clays have high shear strengths. However, high pore water content can reduce the shear strength of any soil.

SOIL WATER

Introduction

The water content of soils is extremely important in determining the properties of the soils. Many of the parameters of groundwater, such as porosity and permeability, have already been considered in Chapter 4. To complete an understanding of the importance of soil water in geotechnics, flow nets and pore pressure are considered below.

Flow Nets

The flow of water through a saturated medium can be represented by flow lines and the different amounts of head can be shown by equipotential lines. The total pattern of flow lines and equipotential lines is called a flow net (Figure 12.5). Note the similarity in concepts with those of the flow of electric currents through the Earth as described in Chapter 6.

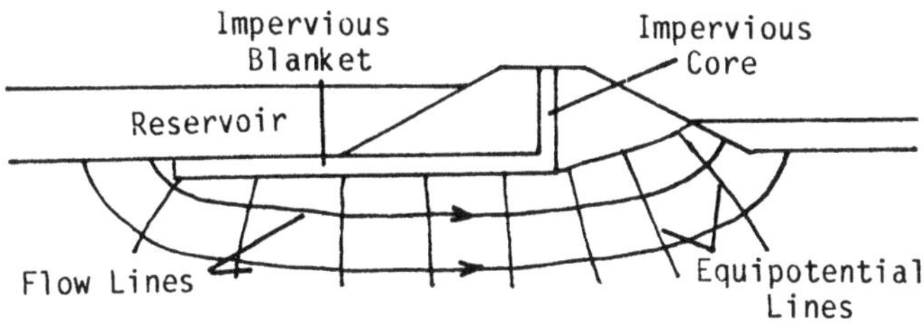

Figure 12.5: The Flow Net Through an Earth Dam.

Pore Pressure

Pore pressure is a combination of the hydraulic head and the application of a compression load. Pore pressure variations are important in determining the stability of a dam. If the pore pressure in fractures and/or bedding planes is increased by an increase in water level in the reservoir, then the shearing strength of the foundation material is decreased and the dam wall may slide down-stream.

GEOTECHNICAL INVESTIGATIONS

Engineering geologists must determine as much as possible about the geology of an area including the physical properties of the rocks and soils before an engineer can fully design the structure desired. Geological mapping can give a great deal of information in many cases, particularly when features such as faults and joints are mapped. If the area is covered by soil, alluvium or glacial material then the task is made considerably more difficult and various other methods of gaining subsurface information are used.

Geophysical surveys, particularly resistivity and refraction seismic techniques, are widely used. These are particularly useful in determining depth to bedrock and water tables - factors which are obviously of vital importance to the civil engineer.

Penetration testing which involves hammering a steel rod into the soil and correlating the number of blows with the depth of penetration is also a useful technique. The greater the number of blows per unit increase in depth corresponds to greater compressibility. Field determinations of shear strength can be made by rotating a metal vane, similar to the feathered end of a dart, in the soil. The shearing strength is directly proportional to the torque required to turn the vane.

The various drilling techniques described in Chapter 8 can be applied to overburden or hardrock investigations. The information obtained is useful not only in determining the type of material passed through but in many cases (such as diamond drilling) can also provide samples of the material in sufficient volume to enable commpressive strength and other properties to be determined. However, the results of even a most comprehensive drilling programme can still provide inconclusive information (Figure 12.6).

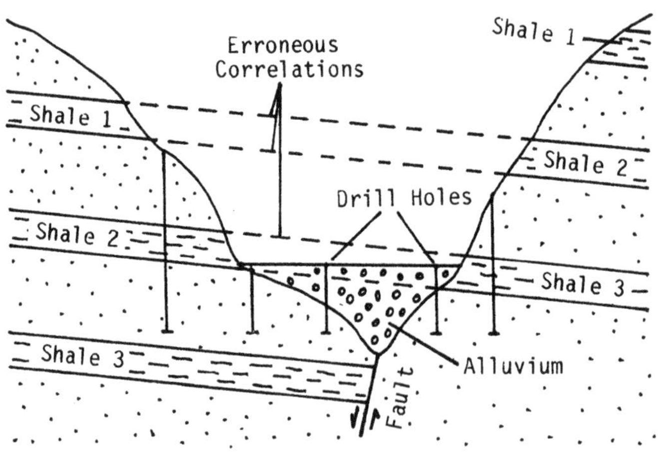

Figure 12.6: Site Investigation for a Dam. Simple Correlation of Strata Implies No Fault.

FOUNDATIONS FOR BUILDINGS

Introduction

Different authors define footings and foundations somewhat differently. In a very general sense, a foundation is the total support of a structure and thus includes the lower part of the structure in contact with earth or rock and also the soil and rock below. This section will consider only the lower parts of structures.

Footings

Footings spread the weight of the building over a larger area to decrease the unit load. The larger the area of the footing the smaller the unit load transmitted to the bearing material. Two basic types of spread footings, individual and continuous are illustrated in Figure 12.7.

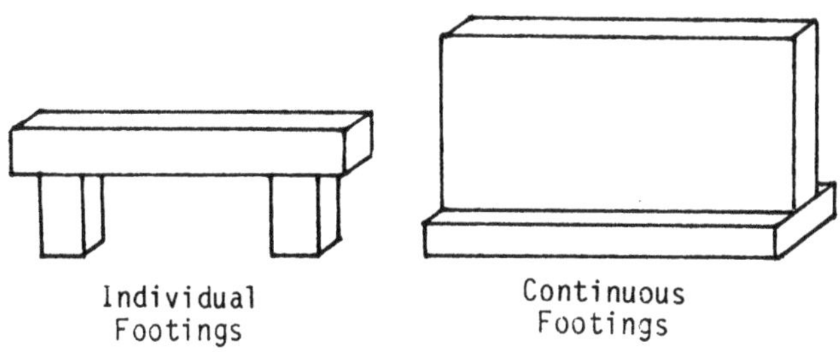

Figure 12.7: Individual and Continuous Footings.

Individual footings may be extended into the soil by drilling a hole and then filling it with reinforced concrete to form a pier. This technique is often used when the soils are expansive clays.

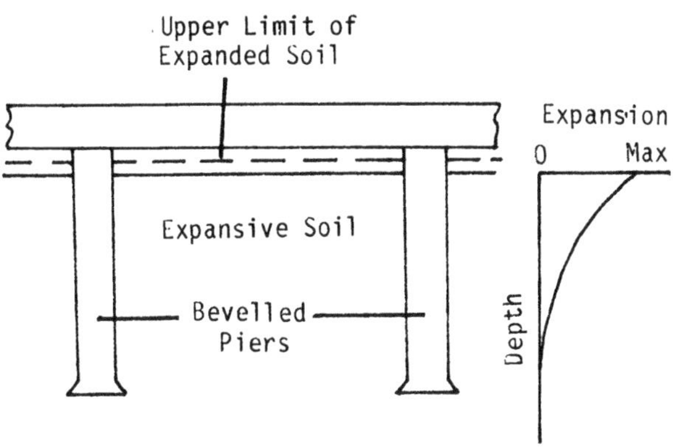

Figure 12.8: Bevelled Pier Footings in Expansive Clays.

The amount of expansion and contraction of the clays decreases with depth. If the base of the pier is below the zone of significant movement then the structure is not likely to fail. In some cases the pier may be bevelled, that is, have an undercut at the base to increase the bearing area (Figure 12.8).

A variation of the spread footing, sometimes considered a foundation by some authors, is the mat or raft. These are usually made of reinforced concrete and extend under the entire structure.

Pile Foundations

Piles may be of timber, concrete or steel and are driven into the ground by a hammer. Piles are used as foundations not only for buildings but also for bridges and similar structures. There are two main types of pile:

1. Friction Piles. The material through which the pile is driven acts as the main support for the pile (Figure 12.9a).
2. Point, or End-Bearing, Piles. The pile is driven until firm, non-plastic material, such as bedrock, supports its tip (Figure 12.9b).

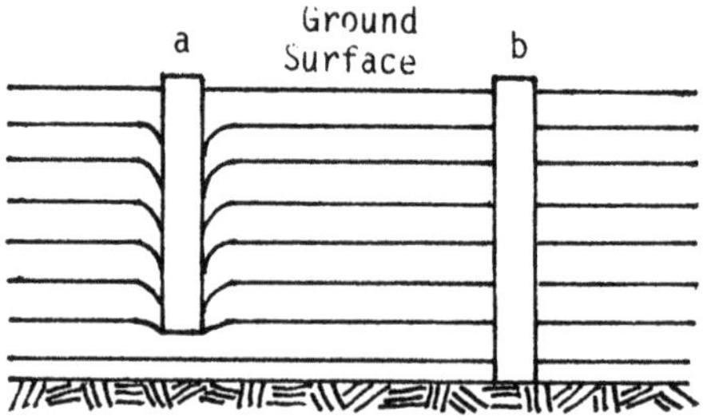

Figure 12.9: Pile Foundations.
(a) Friction Pile.
(b) End-Bearing Pile.

SETTLEMENT

Settlement other than that due to compressibility of soil foundations is mainly due to:

1. Removal of water by pumping or draining. This is commonly due to draining of excavtion sites or, on a longer term basis, the extraction of water or oil from an under-

ground reservoir.

2. Mine workings, both current and abandoned. Although most settlement is gradual and, in the former case at least, can be controlled, in some cases the settlement can be instantaneous with spectacular and catastrophic results.

SLOPE STABILITY

Slopes tend to fail by shearing on a circular arc called the failure or slip surface (Figure 12.10).

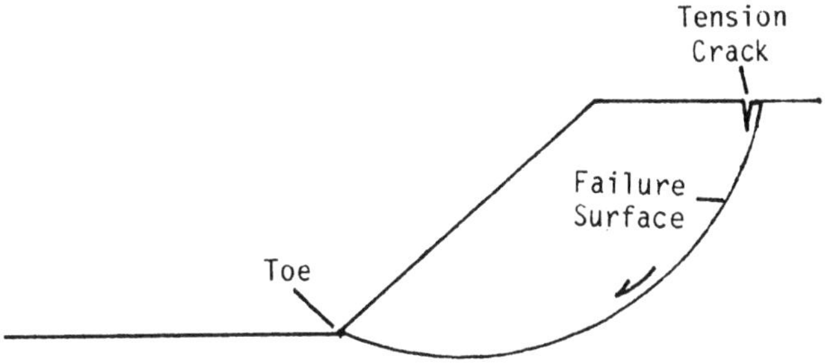

Figure 12.10: The Failure of a Slope.

The shape and position of the failure surface depends on the pore pressure and the shearing strength of the material. In cohesionless materials such as dry sand no failure surface develops and the material simply assumes its angle of repose.

The most common cause of slope failure is an increase in moisture content which decreases cohesion. Not only will soils fail in this way but if there are dipping strata slabs of material may slip. Methods to counteract this problem are:

1. The surface of the slope is sealed and well drained so that no water can penetrate downwards into the material. This can be accomplished by compacting the surface, covering the surface with grass or sprayed concrete and providing sufficient drainage channels.
2. The material is drained internally by drilling holes into sensitive zones thus allowing free drainage and increasing friction (Figure 12.11).
3. The potentially unstable material is physically anchored to prevent slipping.

Alternatively, the slope is prevented from failing by constructing a retaining wall. Retaining walls are positioned at the toe of a slope to oppose the gravitational forces acting on the material. In most cases where a retaining wall is

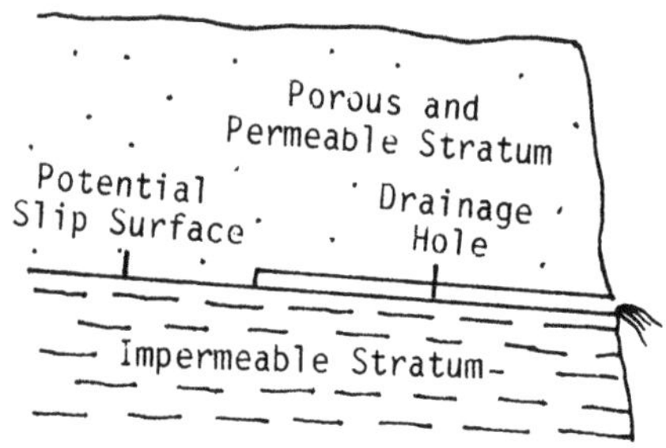

Figure 12.11: Drainage of a Potential Slip Surface.

needed, the toe of the slope has been removed as part of construction work such as in road or rail cuttings. A typical retaining wall is illustrated in Figure 12.12. A number of important conditions

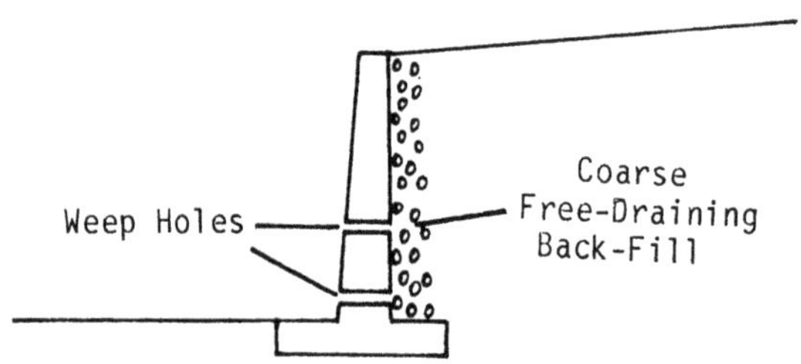

Figure 12.12: A Typical Retaining Wall.

must be met if the retaining wall is not to fail. The wall is acting as a dam and if groundwater is not allowed to freely drain from behind the wall the water pressure is likely to overturn the wall. To allow the water to drain freely, the backfill of the wall should be relatively coarse material which will also have a high angle of rest and put less pressure on the wall. Holes which pass through the wall (weep holes) allow the water to drain away from behind the wall. Alternatively, a lateral drain is incorporated and the water is drained away from the protected area.

An example of a massive retaining wall is illustrated in Figure 12.13. The Folkestone Warren

Figure 12.13: Artificial Toe, Folkestone Warren, England.

is an area of landslides between Folkestone and Dover in southern England. The main London-Dover railway line passes through the Warren and has been subjected to several landslips, some derailing trains. Chalk overlies the Gault Clay, the boundary dipping gently out to sea. Groundwater permeates through the chalk and decreases the shear strength at the chalk-clay boundary resulting in large movements of chalk seawards. After each slide, the transported material formed a toe to the slope and temporarily stopped further slides. However, the action of waves removed this material and further slides occurred. The solution was to construct a massive artificial toe of concrete as shown in Figure 12.13. No further earth movements of significance have occurred since.

The material from which retaining walls are constructed is normally reinforced concrete. In recent years the reinforced earth method of retaining structures has gained wide acceptance. The wall is made up of vertical, interlocking concrete panels. Attached to these panels are strips of metal which run back into the fill material behind the wall. As the wall is built up layer by layer the fill is compacted incorporating the metal strips which then act as anchors for the wall panels (Figure 12.14). One of the main advantages

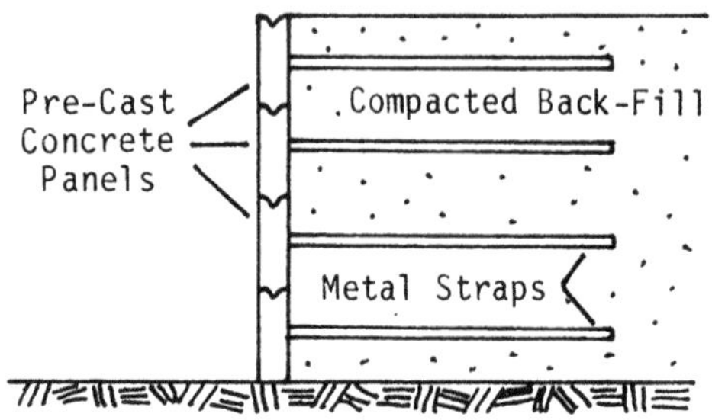

Figure 12.14: Reinforced Earth Wall.

is that vertical walls can be constructed in areas of cohesionless backfill.

Slope stability and associated problems are of great concern to urban developers in Hong Kong. The various solutions to such problems are well documented in the first draft of the geotechnical manual produced by a Hong Kong advisory Steering Group (Anon., 1979).

GENERAL EXCAVATION PROBLEMS

Side-Wall Problems

In an excavation into dipping bedded or foliated rocks there is the possibility of rock slides into the excavation (Figure 12.15). If the excavation

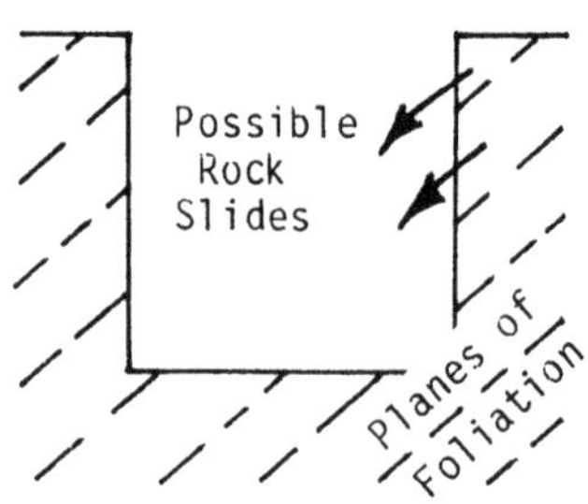

Figure 12.15: Excavation Side-Wall Failure.

is in unconsolidated material such as sand or gravel, then the walls must be properly shored up. This simple requirement is often overlooked, especially during suburban trenching operations for public utility pipes and cables, and there is often consequent loss of life.

Ground Water Problems

If the excavation is deep and/or the water table is near the normal ground surface then some method must be used to remove water from the workings. Simple pumping from the bottom of the excavation is not satisfactory unless a deep sump is used and the material is free draining. A widely used technique is the well point method. A number of wells are drilled in the vicinity of the excavations. When the water is pumped out of the wells the drawdown results in the water-table falling below the lower limits of the excavation (Figure 12.16). This

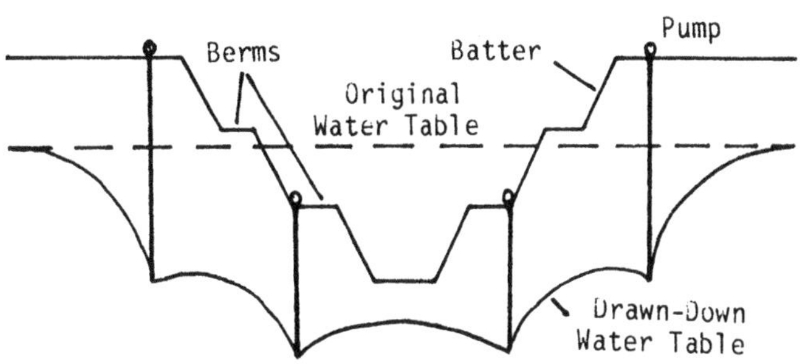

Figure 12.16: Well Point Drainage of an Excavation.

method was used during the excavation of the Bowman's trial coal pit (Figure 10.8).

An alternative method, commonly used in colder countries, is to freeze the material immediately surrounding the excavation.

DAMS

Introduction

The geological investigations needed when determining the siting and design of dams must be complete and detailed. Features such as rock type, structure, weathering, fractures and fissures must all be considered.

The main types of dams are:

1. Earth or rock-fill dams which are constructed of several layers of earth materials and acts as a gravity dam.
2. Concrete gravity dams.
3. Concrete arch dams (Figure12.17a).
4. Buttress dams.

Figure 12.17a: The Myponga Concrete Arch Dam, South Australia.

The main considerations are that the material that the dam rests on must be able to carry the weight of the dam without failure and that it is impervious. In the case of arch dams in particular, the abutments must be impervious and be able to transmit stress without failing. In the case of earth dams the dam must be constructed so that sudden variations in water level will not cause the dam to fail and that the dam is impermeable.

Some geotechnical problems associated with dams are considered below.

Horizontal Forces

These are caused by the lateral pressure of the contained water and the silt which inevitably deposits on the reservoir floor. The gravity dam resists these forces by its mass; the buttress dam by its buttresses and the arch dam by transmitting these forces to the abutments.

Seepage and Leakage

There is always permanent movement of water from the reservoir under, around and through the dam. The "normal" amount of flow is quite acceptable and is called seepage. If this movement of water increases to abnormally high levels this is called leakage. The most obvious concern about leakage is that the dam is simply not doing the job for which it was designed. To prevent leakage in an earth dam, a relatively impervious core is incorporated into the structure. Clay is widely used because of its low permeability and its ability to flex rather than break during movements of the dam wall. More recently, cores have been made of asphalt (Figure 12.17b).

Figure 12.17b: Forming the Asphalt Core of the East Dam of the High Island Reservoir, Hong Kong.

Of more serious concern is that leakage can cause erosion of the dam wall and/or its foundations with possible consequent failure of the dam.

Very detailed foundation investigations must

precede the construction of a dam and if the potential for leakage is indicated then corrective measures must be taken. If the problem is serious, the design of the dam may have to be changed or, in extreme cases, the site abandoned. Invariably, any loose and weathered material must be removed from the dam site. Deeper leakage problems may be corrected by grouting. This is carried out by drilling a series of holes along the axis of the dam and pumping special cement down the holes under pressure. The cement seals off potential leakage channels by forming a grout curtain (Figure 12.18).

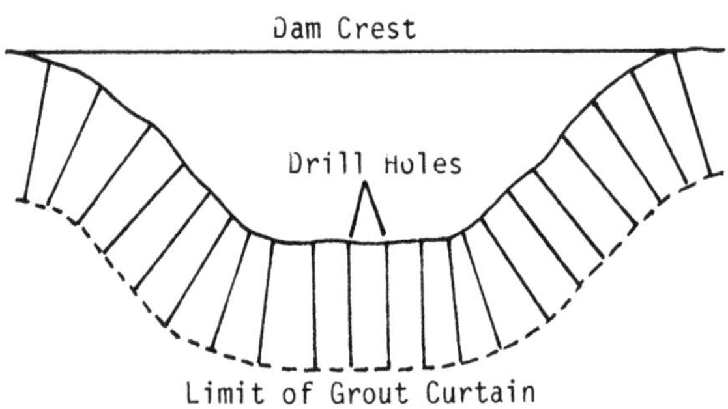

Figure 12.18: A Grout Curtain.

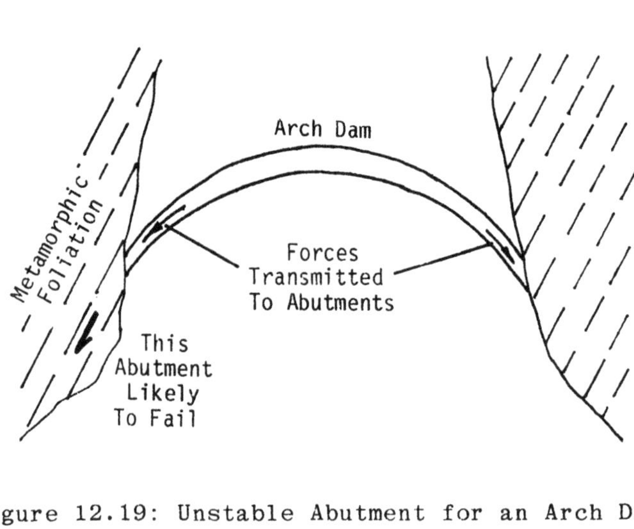

Figure 12.19: Unstable Abutment for an Arch Dam.

Abutment Problems

In the case of arch dams in particular, the geology of the abutments may be a contributing factor in the failure of a dam. Figure 12.19 illustrates the case where the strike of foliation of the abutment rocks is parallel to the main stress direction applied by the arch dam. This relationship can lead to the abutment failing by movement along foliation planes.

Earth and Rock-Fill Dams

These types of dams are built of materials ranging in size from clay to boulders. To prevent leakage a clay, concrete or bituminous asphaltic core is incorporated in the design (Figures 12.5 and 12.17). If a concrete core is used, it is essential that the foundations will not settle differentially which will cause the core to fracture. It is for this reason that most cores in dams of these types now being constructed are either clay or asphaltic. In some cases the upstream side of the dam wall is covered with an impervious layer such as concrete or steel. Not only will this decrease the flow of water through the dam wall but it will decrease the possibility of erosion.

Failure

Dams may fail due to causes other than those mentioned above. These causes may be:

1. Earthquakes (it is surprising to learn of the number of dams that in the past were built in rather sensitive zones).
2. Sudden drop in water level which may result in earth-rock fill dams collapsing by shear failure.
3. Poor protection of the reservoir side of the dam from wave action which will slowly erode earth and rock-fill dams. This problem is usually overcome by covering the dam face with rip-rap. This is a surface of natural rock or concrete blocks which break up the wave action and decrease the energy available for erosion (Figure 12.20).
4. Insufficient spillway capacity. If a spillway is too small, then in times of exceptional flooding water may overtop the dam and erode the face and downstream foundations.

Figure 12.20. Placing Rip-Rap on the West Dam of the High Island Reservoir, Hong Kong.

PAVEMENT FOUNDATIONS

Introduction

Roads or airstrips must distribute wheel weight so that the bearing value of the underlying material is not exceeded. The material upon which a pavement is constructed is called the sub-grade. On top of the sub-grade various layers, or courses, of different material are placed and constitute the pavement proper (Figure 12.21). The pavement must be durable and weatherproof.

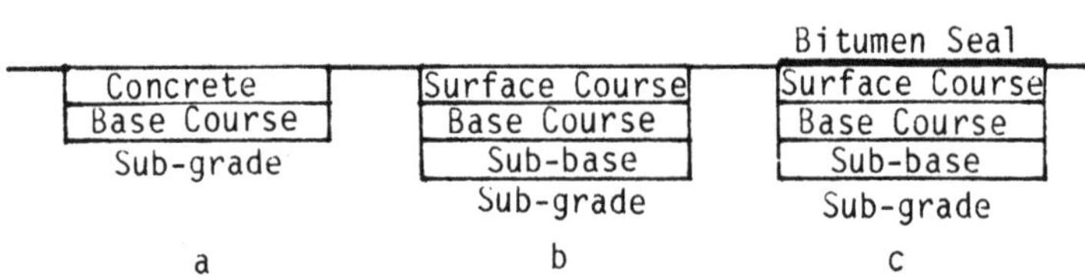

Figure 12.21: The Structure of Pavements.
(a) Rigid Pavement.
(b) Unsealed Flexible Pavement.
(c) Sealed Flexible Pavement.

The Sub-Grade

The physical (and to a lesser extent the chemical) characteristics of the foundation material determine the design of a pavement. The bearing strength determines the depth of excavation that may be needed and the thickness of the subsequent layers. In an extreme case, the nature of the sub-grade is of greatest concern in arctic and subarctic areas. In such cases the foundation is frozen solid in winter but acts as a bog in summer. The special work required includes draining the area in the vicinity of the pavement then placing a mat or raft of timber. The road is then constructed on top of this surface.

The Nature of the Course Materials

This is determined by laboratory investigations to determine size distribution, plastic and liquid limits, shrinkage, compacting properties and bearing strengths. In some cases, where there is no suitable material within an economical transporting distance, then the unsatisfactory material may be upgraded by stabilisation. This is usually carried out by the addition of cement, lime or other chemicals.

The sub-base is incorporated if the sub-grade is of particularly poor bearing quality and is usually the unscreened product of a quarry or pit. The base course is more carefully selected gravel or soil with specific limits of size grading and other soil properties such as liquid and plastic limits. The surface course is the most critical of the layers and has to satisfy exacting size and property limits. These limits are slightly different if the course is sealed or unsealed. The sealing of a road, usually by a bituminous layer, is primarily to prevent water entering the pavement material and thus decreasing its strength.

During construction of the pavement, each course must be compacted so that the material attains a high shear strength, low permeability and high density.

REFERENCES

Anon., 1979. Geotechnical Manual For Slopes. Public Works Department, Hong Kong.

Capper, P. L. and Cassie, W. F., 1960. The Mechanics of Engineering Soils (3rd Ed.). Spon, London.

Krynine, D. P. and Judd, W. R., 1957. Principles of Engineering Geology and Geotechnics. McGraw-Hill, New York.

Legget, R. F., 1962. Geology and Engineering (2nd Ed.). McGraw-Hill, New York.

Peters, W. C., 1978. Exploration and Mining Geology. John Wiley, New York.

Sowers, G. F., 1979. Soil Mechanics and Foundations: Geotechnical Engineering. Macmillan, New York.

FURTHER READING

TEXTS

Ashworth, R., 1972. Highway Engineering. Heinemann, London.

Craig, R. F., 1983. Soil Mechanics (3rd Ed.). Van Nostrand Reinhold (UK), Wokingham.

Knight, M. J., Minty, E. J. and Smith, R. B. (Eds.), 1983. Collected Case Studies in Engineering Geology, Hydrology and Environmental Geology. Engineering Geology Specialist Group, Geol. Soc. Aust., Sydney.

Terzaghi, K. and Peck, R. B., 1967. Soil Mechanics in Engineering Practice (2nd Ed.). John Wiley, New York.

JOURNALS

Journal of the Soil Mechanics and Foundations Division, Proceedings, ASCE.

Chapter Thirteen

ENVIRONMENTAL GEOLOGY

INTRODUCTION

Defining environmental geology is difficult. In the broadest sense geology directly and indirectly influences all parts of the environment. Flawn (1970) defines environmental geology as a "branch of ecology (sic) in that it deals with relationships between man and his geological habitat; it is concerned with the problems that people have in using the Earth - and the reaction of the Earth to that use". Such a definition includes engineering geology and economic geology.

This chapter will consider the "relationships between man and his geological habitat" which have not been covered in other chapters. These include volcanic activity, seismic activity, health, agriculture, pollution and disposal of wastes.

VOLCANIC HAZARDS

Volcanic activity can cause destruction by lava and mud flows and by ash falls (Figure 13.1). The most devastating eruption is the nuée ardente which devastates all in its path as in the case of St. Pierre in Martinique in 1902. Volcanologists are becoming skilled in predicting where, when and how eruptions may occur. However, much more research is necessary to make accurate predictions. For example, in Hawaii, as magma chambers fill beneath volcanoes, the volcanoes actually bulge outwards. Sensitive tiltmeters record such movements and are used to predict eruptions. Seismic activity is also useful because seismic activity tends to increase before eruptions.

A knowledge of the magma type and the previous history of a volcano will give some indication of the style of eruption and the area likely to be affected. Such work resulted in the majority of the National Park around St. Helen's volcano in western U.S.A. to be cleared of people before the major eruption of 1980. Unfortunately, a vertical eruption was predicted instead of the sideways eruption which killed a number of people. The area of the U.S.A. affected by falling ash and dust was also very large.

Figure 13.1: Maori Village Partially Buried by Ash During the 1886 Eruption of Mount Tarawera, New Zealand.

Advances in technology can increase the range of problems associated with volcanic activity. In mid-1982, several international jet airliners experienced engine failures when flying through a cloud of ash and dust resulting from eruptions in Java.

SEISMIC HAZARDS

Damage and deaths caused by earth movements are more newsworthy events because earthquakes generally affect much larger areas than volcanic eruptions (few cities are built on the slopes of volcanoes!).

Destruction can result from:

1. Partial or total collapse of buildings, dams and other structures (Figure 13.2).
2. Landslides and avalanches triggered by earthquakes.
3. Tsunamis (or tidal waves).

Figure 13.2: Public Building Badly Damaged by the 1978 Mekering Earthquake, Western Australia.

Areas of the Earth's crust which are unstable are relatively well known because of historic records. If populations refuse to leave a high risk area for various reasons, then the prediction of earthquakes and ways of decreasing their effects is important. A great deal of work is being carried out throughout the world in attempts to accurately predict earthquake activity. There have been some successes mainly as a result of:

1. Measurements of strain in rocks.
2. Measurements of dissolved gases in groundwater.
3. Measurements of variations in electrical propertis of rocks.
4. Observations of water level changes in wells and bores.
5. Observations of animal behaviour.

Even if earthquake activity can be predicted and a population evacuated from the affected area, there is much pressure to ensure that property damage is kept to a minimum. This involves geological mapping to locate active faults and then formulating town planning regulations which specify the kind of land use and types of construction methods in relationship to the risk areas.

HEALTH

The Natural Environment

There is a growing body of data which suggests that there is often a close relationship between the natural trace element concentration in the rocks, soils and waters of a region and the health of the human population of that region (Cannon and Hopps, 1971).

The effects of trace elements which occur in any organism depend on the concentration or dose of the elements. There is an optimum concentration range which is beneficial to life. The beneficial effects decrease as the concentration increases or decreases away from the optimum until the concentrations are harmful (Figure 13.3).

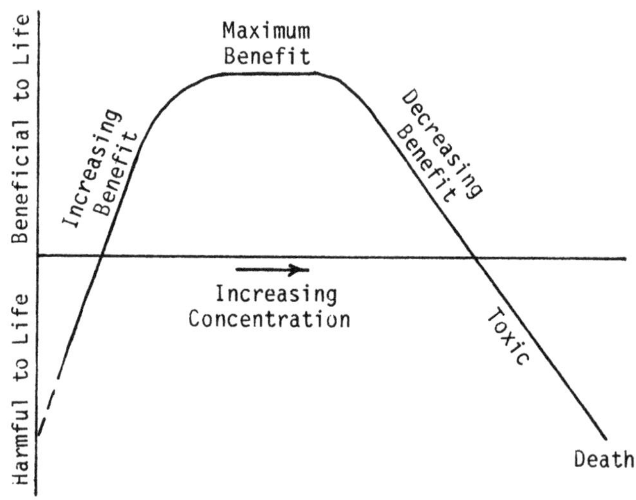

Figure 13.3: A Generalised Dose-Response Curve. (After Keller, 1976).

One of the best known of the effects of trace element concentration on health is that of fluorine. In many parts of the world it was noted that populations had a "less than average" incidence of dental caries. This lower incidence was traced to higher than average concentrations of fluorine (in the form of soluble compounds) in the geological environment, particularly water used for drinking. This relationship has led to the widespread practice of fluoridising drinking water in other areas.

Conversely, iodine deficiency causes goitre which results in the enlargement of the thyroid gland in the neck. Cretins, or mentally retarded dwarves, are the result of iodine deficiency in the mother during pregnancy. Some breast cancer is also related to iodine deficiency.

Environmental Pollution

Civilisation has always affected the environment, usually to the detriment of both. The most obvious effects such as pollution of waterways with "normal" residential and industrial wastes are rightly considered undesirable and the effects of such pollution on a population are nowadays minimised.

More insidious is the addition of trace elements to the environment in quantities which are toxic to human and animal populations. Often, these trace elements find their way into organisms by a chain of events rather than directly. This makes the determination of the cause of the toxic effects more difficult. One of the earliest examples is the addition of toxic levels of lead into the drinking water of Roman cities due to the ues of lead pipes. In the opinion of some authorities, this effect was one of the major causes of the fall of the Roman Empire. In Great Britain recent surveys have found similar high lead levels in some drinking water for the same reason although it is doubtful that the decline of the British Empire is the result!

An example of a complex path of pollution is the mercury poisoning of the people of Minimata Bay in Japan. The mercury was contained in industrial waste which was released into Minimata Bay. The mercury passed into the fish cycle and the fish were subsequently eaten by the local population with tragic results.

An increasing concern about trace element pollution associated with major highways has resulted in many investigations into the lead distribution in adjacent soils (for example, Bottomley and Boujos, 1975). Cadmium is also present in high levels near highways, the source of the metal being vehicle tyres (Hosie and others, 1978). Such results have major implications for siting residential and school buildings and for growing vegetables for human consumption.

The detection of trace elements in environments such as those described above is based on standard geochemical techniques. Not only are biological materials analysed but also waters and soils (Thornton, 1974; Thornton and Webb, 1974).

AGRICULTURE

There has recently been a growing awareness of the importance of trace elements in the production of crops and livestock. Most people are aware that soils which are subjected to intensive cultivation require periodic addition of fertilisers of different types to replace material used in the production of crops. The most common fertilisers are superphosphates and nitrogenous and organic fertilisers.

Recent studies have shown the importance of trace elements in agriculture. Crops grown on soils developed over ultrabasic rocks, for example, contain high levels of nickel and cobalt. Excessive molybdenum intake by sheep results in copper deficiencies with consequent low weight. Soils developed over sands contain low cobalt and this deficiency too is detrimental to sheep.

Toxic levels of trace elements are difficult to counteract. However, deficiency problems can be solved by adding the relevant trace element(s) to the soil or directly to the animal in feed.

Areas affected by past and present mining activity often have toxic levels of lead, arsenic, copper, zinc and cadmium and care must be taken during the production of livestock and crops for human consumption.

DISPOSAL OF WASTES

Traditional disposal of wastes by dumping in waterways and oceans is being increasingly opposed by modern civilisation. Dumping of solid refuse in land reclamation areas is generally accepted although contamination of groundwater by seepage is a constant problem.

One of the most controversial waste disposal problems of the late twentieth century is the disposal of wastes from the nuclear fuel cycle.

The disposal of low-level and medium level

radioactive wastes is now a routine industrial operation. Such disposal must have careful design and subsequent monitoring. For example, low-level wastes from the mining and milling of uranium ore can be effectively isolated from the environment by shallow burial with a minimum of one metre of compacted soil (Figure 13.4).

Figure 13.4: Low-Level Radioactive Waste Dump, Radium Hill, South Australia.

The disposal of high-level wastes requires more careful consideration. Most of these wastes, in liquid form, are stored in large containers at the Earth's surface. Such containers are likely to, and have, failed with subsequent local pollution. The present direction of investigations to solve this problem is to determine the best method of disposing these wastes deep in the Earth's crust.

In the first of two main proposals, the wastes, as liquids or solids, are sealed in drums and placed in island-arc trenches where the sedimentation rate is very high. The expected result is that the wastes will be rapidly buried and ultimately they will be involved in the subduction processes of convergent plate margins. It is supposed that the wastes will be either subducted into the mantle or incorporated into the metamorphosed trench sediments which ultimately form new mountain ranges. One suggested site is the subduction zone to the east of Japan so that nation can dispose of the large quantities of the waste resulting from its large commitment to a nuclear power system.

The principal argument against this proposal is that there is insufficient knowledge about the integrity of the containers during the burial stage of the process.

In the second proposal, the wastes, as solids formed by the vitrification or synroc processes (Ringwood, 1978), are sealed in suitable containers. These containers will be designed to be resistant to corrosive groundwater and also to react to stress by deformation rather than by fracture. The containers will be placed in the crust by either drilling a large diameter hole several kilometres deep, lowering the containers into the hole and sealing the hole with clay and concrete; or using normal mining operations to prepare chambers for the containers at depth.

The type of geological materials suitable for such disposal sites include granite, salt and clay. Each of these types have, or are associated with, some favourable properties such as impermeability, geological stability, lack of groundwater and reaction to stress by deformation rather than by fracture.

Sweden is currently testing Precambrian granites to determine if the mining option is feasible. So far the results have been favourable. West Germany is considering clay deposits and Great Britain is considering drill holes into granites.

REFERENCES

Bottomley, G. A., and Boujos, L. P., 1975. Lead in Soil of Heirisson Island, Western Australia. Search, **6**: 389-390.

Cannon, H. L. and Hopps, H. C. (Eds.), 1971. Environmental Geochemistry in Health and Disease. American Association for Advancement of Science Symposium, Dallas, Texas, 1968. Geol. Soc. Amer. Memoir 123.

Flawn, P. T., 1970. Environmental Geology. Harper and Row, New York.

Hemphill, D. D. (Ed.), 1974. Trace Element Substances in Environmental Health - VII, 1974. A Symposium. Univ. Missouri, Columbia.

Hosie, D. J., Bogoias, A., De Laeter, J. R. and Rosman, K. J. R., 1978. The Cadmium Content in Soils at Heirisson Island, Western Australia. Search, **9**: 47-49.

Keller, E. A., 1976. Environmental Geology. Merrill, Columbus

Ringwood, A. E., 1978. Safe Disposal of High level Nuclear Reactor Wastes: A New Strategy. Australian National Univ. Press, Canberra.

Thornton, I., 1974. Applied Geochemistry in Relation to Mining and the Environment. In Minerals and the Environment, 1974. A Symposium, London, Inst. Min. Metall., London.

Thornton, I. and Webb, J. S., 1974. Environmental Geochemistry: Some Recent Studies in the United Kingdom. In Hemphill, D. D. (Ed.), 1974: 83-87.

FURTHER READING

TEXTS

Tank, R. (Ed.), 1973. Focus on Environmental Geology. Oxford University Press, London.

JOURNALS

Australian Journal of Agricultural Research
Environmental Health
Soil Science

Chapter Fourteen

GEOLOGY, ECONOMICS AND POLITICS

INTRODUCTION

The survival of our industrial civilisation depends on reliable supplies of metals, non-metals and fuels. To ensure this, not only must time, effort and money be invested in locating and extracting Earth resources, but reliability of continued supply must also be ensured. Some of these problems are considered in this chapter. Apart from the few references given here, one of the most prolific and stimulating sources of information on this topic is the daily press, particularly the financial pages.

QUANTITY OF EARTH RESOURCES

In Chapter 1, the term "resource" was considered to be a very broad definition of the mineral content of the Earth. Thus, although resources do have a finite limit in global terms, in terms of the life of our civilisation, they are sufficient for the foreseeable future. The major proviso is that unless similar grade deposits to those being currently mined can be found deeper in the Earth's crust then lower grade deposits will have to be extracted and treated. In either case, this implies that the consumer will ultimately have to pay correspondingly more for the product.

Govett (1978) suggests that there will be a lower limit to the grade of some types of deposits (Figure 14.1). The principal reason is the exponential increase in energy requirements to produce metal from the decreasing grades of ore using currently conventional mining and metallurgical techniques. However, it has already been noted in Chapters 10 and 11 that the developments in mining and mineral processing of using fluids of various types (solution mining, leaching, microbiological mining and so on) can increase the ease of mining operations and also make lower grades economic propositions.

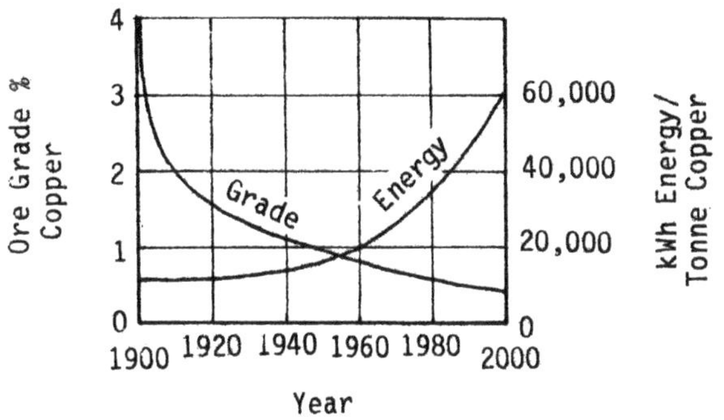

Figure 14.1: Ore Grade and Energy Requirements in Copper Production. (After Govett, 1978).

The problem for petroleum resources appears to be more critical. However, it is interesting to note that in spite of constant predictions of an oil shortage no major natural cause of shortage has yet occurred. This may be because it is of no financial benefit to a company to spend today's money to search for more and more deposits if there is no demand for the product. The company would have valuable capital tied up (in terms of the cost of exploration) in a non-producing asset when the money could be at least earning interest in investments and producing a real return. The critical time in the petroleum industry appears to be about twenty years, that is, there always seems to be about twenty years' reserves of petroleum, world wide, at any point in history.

GROWTH RATE OF THE DEMAND FOR EARTH RESOURCES

The demand for minerals, fuels and their derived products depends on a number of inter-related factors. Although subject to minor fluctuations, the growth in the world's average Gross National Product is about five per cent per annum. The other main factor is the continuing increase in the world's population. Both these factors should result in increasing demand for Earth resources even if the standard of living remains constant, which is highly unlikely (it is likely to rise). It is estimated that the per capita demand for minerals will grow at about four to five per cent per annum (Govett 1978).

Minor fluctuations in this demand are the result of such events as wars, provision of aid to lesser developed countries (usually to open up new markets for the donor nation) and expansion of, and changes in, technology. For example, tin-plate is now manufactured by electrolytic methods rather than by the hot dip method leading to a 60 per cent drop in the demand for tin. The use of titanium in the aerospace industry is an example of how new technology increases the demand for previously little used metals.

To a large extent, free market forces can cope with these variations. If prices reflect the supply-demand situation, then when a metal or fuel is in short supply its price will tend to rise. If the rise is too great to be absorbed by the consumer a substitute material may be used. For example, aluminium is substituting for copper in electricity power cables and for tin-plate in the canning industry. In a similar fashion, after the rapid increase in the price of petroleum in the early 1970s, many countries returned to coal as a power source, or increased the production of electricity by nuclear power plants. Some even considered all possible energy sources before making more conventional choices (Inhaber, 1978).

Environmental issues also affect the supply-demand situation and hence the price of a commodity. For example, in the mid-1970s, base metal smelters were required to decrease the amount of atmospheric pollution levels, particularly sulfur dioxide emissions. In many cases, it was more economic to construct new smelters to conform with the new regulations. This resulted in a considerable drop in smelting capacity from the time the old smelters were closed until the new smelters were constructed. A tightening of supply resulted in increased prices.

The large price rises in petroleum products in the early 1970s, and the embargo of exports by the OPEC countries to those countries which did not support their stance in the Middle East resulted in many countries changing their political views to ensure continued oil supplies. Most countries came to realise that it was not wise to depend on a small group of producers for their energy needs.

Exploration for petroleum expanded in non-OPEC countries, measures were taken to conserve energy consumption and alternative energy supplies were used as noted above.

The higher energy costs led to a general world recession with consequent lowering of demand for petroleum. As a result the price of petroleum in dollar terms and hence in real terms is currently falling. Many of the petro-dollars accumulated by OPEC were invested in western nations from where they were re-invested in the lesser developed nations, particularly in mineral development projects. As the world moved into recession, however, the demand for metals fell. Normally, this would result in reducing mine and smelter production to keep prices stable. However, because the developing countries were required to repay their loans in foreign currencies (usually the US dollar) production was continued to earn the required foreign exchange and world prices fell even further. The result is that mining companies are finding it difficult to get a reasonable return on their investments in the early 1980s (Figure 14.2).

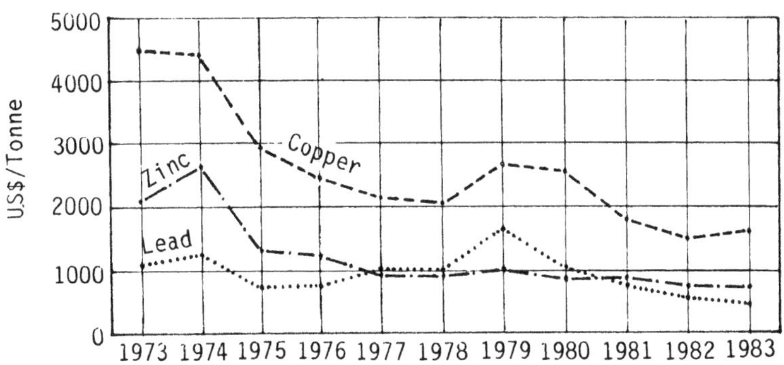

Figure 14.2: Copper, Lead and Zinc Prices Deflated by US Wholesale Price Index, Base December,1983. Data From Wellesley-Wood, 1984.

Prices of metals can vary considerably over relatively short periods. These variations are the result of factors such as strikes, wars, currency exchange rates and political change. Figure 14.3 illustrates the variations in the weekly copper price on the London Metal Exchange (LME) for 1983.

The rapid rise in prices in the early part of

the year was mainly due to the fact that the world considered the pound Sterling to be a petro-

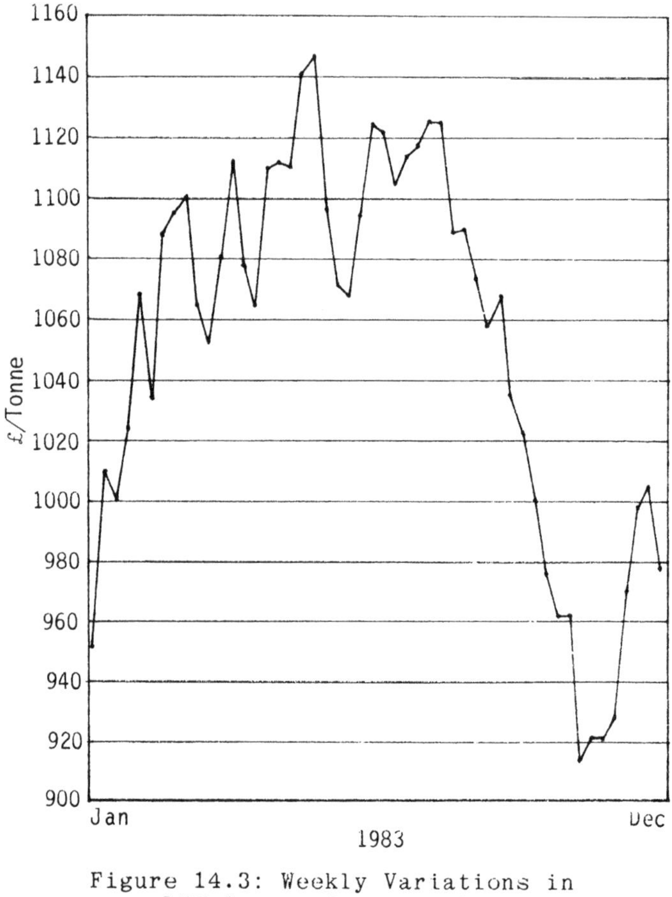

Figure 14.3: Weekly Variations in LME Copper Prices, 1983. Data From the Mining Journal.

currency because of the North Sea oil income. However, during the early part of 1983, the OPEC nations were having difficulty in maintaining high oil prices. As a consequence the pound fell relative to the US dollar and copper prices correspondingly rose. Other factors contributing to this rise were the continuation of industrial action in Peruvian copper mines and increased economic activity in the USA.

The slight fall in the middle of the year was

due to the increased value of the pound but increased car sales in the USA and the threat of strikes in Chilean copper mines led to a recovery in prices.

The long slide in prices in the latter half of the year was due to the pound rising in value, weak demand for metals because of the generally poor economic performance of the world, and overproduction by the lesser developed countries leading to higher copper stocks. The slight rise at the end of the year was due to a slight weakening of the pound.

Such large variations in metal prices make it very difficult for consumers to plan in advance for even relatively short periods of time. There are several methods of evening out the variations in price such as hedging against future variations (at a price). As a result copper producers in Australia and other countries price their product on a moving average of the LME prices.

CONCLUSIONS

The person in the street is affected by all the above situations, because it is the consumer who ultimately pays. Understanding the reasons for variations in the prices of Earth materials may make the consumer more able to make informed decisions in such areas as demands for environmental reforms, trade sanctions and voting choice.

REFERENCES

Govett, G., 1978. Geology - The "Cinderella" Science. Univ. New South Wales Quarterly, (11): 12-13.

Inhaber, H., 1978. Is Solar Power More Dangerous Than Nuclear? New Scientist: 444-446.

Wellesley-Wood, M., 1984. Little Improvement. In Mining Annual Review, Mining Journal, London: 7-17.

FURTHER READING

TEXTS

Al-Chalabi, F. J., 1980. OPEC and the International Oil Industry. Oxford University Press, Oxford.

McDivitt, J. F. and Manners, G., 1974. Minerals and Men (Revised Ed.). John Hopkins Univ. Press., Baltimore.

JOURNALS

Mining Journal
Mining Magazine
Mining Annual Review
The Daily Press

INDEX

Note: Page numbers in bold type refer to figures.

AAS 109, 127
abutments 192, **194**, 195
accuracy 108, **109**
acetate peels 127
acid attack 108
acid rain 19
acidation 51
activators 173
adit **148**
aeration, zone of **43**
agglomeration 160, 164
agriculture 204
albite 9
alginite 23, **33**
alluvial deposits 10, **10**
alpha particles 93
alteration: wall rock, 8
American Petroleum Institute units 130
analytical methods 107-109
anomaly: Bouguer 65; geochemical **98**, 99, 100, **100**, **102**; gravity **66**; magnetic **73**
anthracite **24**
anthraxylon 23, 24
API units 130
apparent resistivity 81, 83
aquiclude 43, **44**, **45**, 46
aquifer 41, **44**, 46; confined, 44, **45**; unconfined, **45**
arch dam 192, **192**, **194**, 195
Arinteiro copper mine, **163**
aromatic petroleum 28, **29**, **30**, 31
artesian: basin 41; flow 44; well, **45**
ash 19, 21
ash falls 199, **200**
asphalt 28
Athabasca Oil Sands 36
atomic absorption spectrophotometry 109, 127
Atterberg Limits: consistency limits 180, **181**; liquid limit 181, **181**, 197; plastic limit 181, **181**, 197; plasticity index 181; shrinkage limit 181, **181**
attritus 23
auger drill 112, 117, **117**
autochthonous coals 21
autogenous mills 162
avalanches 201
average grade 139

back casting 142, **143**

background **98**, 99, 100, **100**
bailer 114
Baku 36
ball mill 162, **162**, **163**
banded iron formation 11, 58, 140
baryte 118
base course 196, **196**, 197
basin: artesian 41
Bass Strait 21
bastnaesite 6
batter **191**
bauxite 11, **12**
beach sand deposits 11, 166, 167, 169
benches 143-145, **143**
bentonite 118
benzene 28
berm **191**
beta particles 93
biotite 8
bit: drill **113**, 119, **120**
bituminous coals 23, **24**
black shale **26**
bleaching 8
blowout 118
blowout preventer **113**, 118
boghead coal 25, **26**, 31
bone coal **26**
bornite 13
Bouguer: effect 64; anomaly 65, **65**, **66**
Bowman's Pit **144**, 191
briquetting 164
British Empire 203
Broken Hill, NSW 170
brown coal 21
bucket excavator **143**
building stone 11, 144
Bushveld Complex 5, **5**
buttress dam 192

cable drill 112, 114, 115, **115**
cadmium 203
calcite 9, 25
cancer 203
cannel coal 21, **26**, **26**
cap rock **34**
carbon-in-pulp 155, 175
carbon isotopes 106
carotene 30
casing: drill **113**, 114, **115**, **116**, 122
catagenesis 32
cell: froth flotation **171**
cement 11
chalcopyrite 9
chargeability 89
China clay 8
chlorite 9
chlorophyll 29, 30
choke **113**, 119
chromite 5, **5**
churn drill - <u>see</u> cable drill
clarain 24, 25
classification: mineral deposits 1; coal 25; crude oils **30**
classifying 160
clay: China 8
coal: allochthonous 21; anthracite **24**; ash 19; autochthonous 21; bituminous 23, **24**; boghead 25, **26**, **29**; bone **26**; brown 21; cannel 21, **26**; classification 25; cyclothem 20, **20**; humic 23; lignite 23; lithotypes 23; macerals 23, 33; measures 20; meta-anthracite 24; mining **138**, 142, 144, **144**; peat 22; rank 19, **24**, 25; seam 20, **20**; semianthracite **24**; series **24**, 25; tectonic setting **15**
coalification 22
coefficient of permeability 46
cohesionless soils 182, 187
cohesive soils 182
collar: drill **113**, 118
collectors 171-173, **172**, **173**

collinite 23, **26**
colorimetry 108
comminution 160
compaction 182
complex pegmatitic deposits 6
compressibility 181, 182, **182**, 186
compressive: deformation 182; load 183; strength 179, **180**
concentrate **175-176**
concentration: criteria, 167; ratio of 177 see also sorting
conductivity: 80, 81, 90; ionic 81; electronic 81
cone crusher 161
confined aquifer 44, **45**
conical pit **144**, 145
connate water 41
consistency limits 180, **181**
consolidation trickling 168, **168**
contact angle 172, 173, **172-173**
contact resistance 81
continental drift 16, 20
continuous footings 185, **185**
continuous velocity logs 130
contrast 101, 103
Cooper Basin 31
copper 7: mining 154 **154** prices, **211**, **212** production costs, **209**
core 120, **120**: barrel 120, 120, 121; catchers 120, **120**; logging 126, 127
core: impervious dam 193, **192**, 195
Cornwall: China clay 8; mine dumps **155**; mineral zoning 7, **7**; tin and copper mines 7, **3**, 135
cost: exploration 52, **53**
covellite 13
cretins 203
critical angle 75, **76**
critical distance 78
cross-cut **148**
crown block **113**
crude oil 28, 29, **30**
crushers: cone 161, gyratory 160; hammer mill 161; jaw 161, **161**; primary, 160
crushing 160
cumulus layers 5, **5**
Curie point 71
cut and fill mining 149, 150, **150**
cut-off grade 139
cutinite 23
cuttings: drill 118, 121; logging 125, 126
cycle: geochemical **95**; hydrologic **40**
cyclone 116, **116**
cyclothem 20, **20**

dams 191-196, **192-194**, **196**; arch 192, **192**, **194**, 195; buttress 192; earth 192, **192**, 195, **196**; failure 195; gravity 192; rock-fill - see earth
declines 149
deformation: compressive, 182; elastic 182; plastic 182
density logs 130
dental caries 203
depressants 173
derrick **113**
development: well 49, **49**, 50
deviating 151
dewatering 174
dielectric constant 81
diagenesis 33
diamagnetic 70, 165
diamond drill 112, 119-122, **121**
diamonds 11, 14, 140
differential settlement 182

dilution: ore 139
dip: magnetic 72
dispersion 96, 98, **97**; train 103, **104**
disposal of wastes 204-206, **205**
diurnal magnetic variation 72
dose-response curve **202**
downcast shaft **148**
drainage 197, 188, **188**, **188**; well point 191, **191**
drawdown 49, **49**, **49**
draw-works **113**
dredging 140, 141, **141**
drift **148**
drill: auger 112, 117, **117**; bit **113**, 119, **120**; cable 112, 114, 115, **115**; casing **113**, 114, **115**, **116**, 122; churn - see cable; collar **113**, 118; core 120, **120**; cuttings 118, 121; diamond 112, 119-122, **121**; dual tube 122; hammer 115, 116, **116**; percussion 112; pipe **113**; reverse circulation 122; reverse flow 122; rods **113**; rotary 112, **113**; ships 118; string 114; tools **112**, wire-line 120; - see also rigs, offshore
drilling: electric 84; exploration 59, 60; mud 118
drill-stem test 119
drives 146, 147, **147**
dual tube drilling 122
dumps: mine **155**, 156
durain 24, 25, **26**

earth dam 192, **192**, 195, **196**
earth wall: reinforced 189, 190, **190**
Earth's magnetic field 71, **71**
earthquakes 195, 200-202, **201**
effective porosity 42
elastic deformation 182
electric drilling 84
electric logs 129-130
electrical sorting: electrostatic 166, **166**; electrodynamic 166, 167, **167**
electrical surveys 80-92
electrochemical activity 81
electromagnetic surveys 80, 89-91, **89-91**
elevation head 44, **45**
emission spectroscopy 109, 127
end-bearing piles 186, **186**
engineering geology 179
enrichment ratio 177
environment 155, **155**, **156**, 156
environmental geology 199
epidote 9
epigenetic mineralisation 3
epithermal deposits 10
epoch: Metallogenetic 8
equipotential surveys 80, 82, **82**
ethane 30
evaporation **15**
evaporites **40**
excavations: groundwater 191, **191**; slides 190, **190**
exinite 23, **26**, **26**, **33**
exploration: cost 52, **53**; drilling 59, 60; reconnaissance 56
extraction: primary 37; secondary 37; tertiary 37

failure surface 187, **187**
farm in 55
farm out 55

feldspar: potassium 8
ferromagnetic 165
field: petroleum **34**, 36
first arrival seismic waves 75
flare pit **113**
flights: auger 117, **117**
Flinders Ranges 14, 103
flint mines 135
flow lines 183, **183**
flow nets 183, **183**
flow sheet **176**
flowing film concentration 168-170, **169-171**
fluid extraction 150-153, **151**, **152**
fluorine 202
footings 184-186, **185-186**; continuous 185, **185**; individual 185, **185**; mat 186; pier 185, **185**; raft 186
Folkstone Warren, England 189, **189**
formation factor 129
formation water 41
fossil magnetism 71
foundations - see footings
fracturing: hydraulic 51, 152, **152**
Frasch process 153, **153**
free air effect 64
friction pile 186, **186**
froth flotation 170-174, **171-174**; activators 173; cells **163**; collectors 171-173, **172**, **173**; depressants 173; frothers 173; roughers 174, **175**; scavengers 174, **175**
frothers 173
fusain 24, 25
fusinite 23
fusion 108

gal 63
gamma 70
gamma logs 130, **131**
gamma-gamma logs 130
gamma rays 93
gangue 160, **161**
gas **29**, **33**, 34
gas detectors 128
gas repressurising 153
gasoline 31, **33**, 34
gauss 70
Geiger counter 93
geochemical: contrast 101; cycle **95**; logging 127, 128; mobility 96, 97, **96**; provinces 99, 103; surveys 58, 59, 101-106
geophone 76
geophysical logging 128-133
geophysical surveys 58,59
geotechnical investigations 183, 184, **184**
geotechnics 170
geothermal gradient 32
ghosts: seismic 80
giant 140
goitre 203
gold 10, 11, 140, 141, 175
gossan 13, **13**
grade 138-140, 208, 209, **209**; average 139; cut-off 139; weighted average 139
gradient: geothermal 32; potential 81; regional gravity 65, **66**
graphite **24**
gravel 11, 141
gravimeters 63
gravimetry 108
gravity dam 192
gravity sorting 167-170, **168-170**; consolidation trickling 168, **168**; flowing film 168-170, **169-171**; jigs 167, 168, **168**
gravity survey 63
gravity survey interpretations 66-69, **66**, **68**
gravity unit 63

grinding 160-164, **162-163**
see also mills
grizzlies 149, **149**, **150**, 160
Gross National Product 209
groundwater **40**, 40; in excavations 191, **191**; locating 47; pressure 44
grout curtain 194, **194**
gypsum 11, 19
gyratory crusher 160

halite 11
halogens 8
hammer drill 115, 116, **116**
hammer mill 161
Hawaiian Islands 41, 199
hazards: seismic 200-202, **201**; volcanic 199, 200, **200**
H/C ratio 32, **33**
head: elevation 44, **45**; hydraulic 44; pressure **45**; total 44
health 202-204, **202**
heavy mineral deposits 11
hematite 11
High Island Reservoir **193**, **196**
Hong Kong 190, **193**, **196**
horizontal integration 54
humic coal 23
Humphries spiral 169, 170, **170**
humus, 21
Hutton, James 4
hydraulic fracturing 51
hydraulic gradient 44
hydraulic head 44
hydraulicking 140
hydrocarbon 28
hydrogen/carbon ratio 32
hydrologic cycle 40, **40**
hydrothermal deposits 6, 8
hypothermal deposits 10

ilmenite 11, 70, 166
impervious core, dam 193, **193**, 195
in situ leaching 153, 154, **154**
inclination 72
incident waves 75
individual footings 185, **185**
induced polarisation surveys 80, 86-89, **86-88**; frequency variation 87, 88; frequency domain 87, 88; time domain 88, 89; transient 88, 89
induction log 130
infiltration **40**
injection: sand 51
intake beds 41, **45**
integration: horizontal 54; vertical 54
intermittent saturation 43, **43**
International Gravity Formula 64
interpretation: gravity surveys 66-69; magnetic surveys 72-74; seismic surveys 76-80
interval transit time 131
iodine deficiency 203
IP - see induced polarisation
iron 7
isotope surveys 106

jack-up drilling rigs 118
Java 200
jaw crusher 161, **161**
jetting 50
jigs 167, 168, **168**
Johnson Well Screen 50, **50**
Joplin lead 106
juvenile water 41

Kambalda nickel deposit 6
kelly **113**; hose **113**

kerogen 31, 32, **33**, 33, 36
kerosene 28
kimberlite 14, 56
Kuroko deposits 11, **15**, 145, **146**

lag 125
landslide 189, 201
lateral log 129
laterolog 129
lateritisation 11
lava flows 199
leach mining 135, 153, 154, **154**
lead 7
lead isotopes 106
leakage 193
Leigh Creek coal mining 142
level **148**, **149**, 149
liberation 159
liquefied petroleum gas 28
lignite 23, **24**
limestones 11
limonite 11
Lindgren 3
liquid limit 181, **181**, 197
lithotypes 23
LME - see London Metal Exchange
logging: continuous velocity 130; core 126, 127; cuttings 125, 126; electric 129, 130; geochemical 127, 128; geophysical 128-133, **131**; magnetic 132, **132**; radiometric 130; sonic, 130
London Metal Exchange 211-213
LPG 28

macerals 23, 33
magmatic deposits 5
magmatic water 41
magnetic: dip 72; field 70, **70**, force 69; inclination, 72; induction 70; logs 132, **132**; poles 69; storms 72; survey 69; susceptibility 70, 132; variation 72
magnetic sorting: low intensity 165, **165**, **166**; high intensity 165, 166
magnetic survey interpretation 72-74, **74-74**
magnetism: diamagnetic 70, 165 ferromagnetic 165; fossil 71; paramagnetic 70, 165; residual 70
magnetite 11, 70
manganese nodules **15**, 141
massive sulfides **15**, **138**
mast: drill **113**
mat footing 186
measures 20
mechanical miner 146, **147**
megaspores 23
Mekering, WA **201**
mercury 203
Merensky Reef 5, **5**
mesothermal deposits 10
meta-anthracite **24**
metagenesis 32
metal detectors 90
metal factor 88
metal prices **211**, **212**, 211-213
metallic polarisation 86, 87
metallogenetic: epoch 8; province 8
metallurgical testing 60
meteoric water 13, 41
methane 28, 29, 35
mica 6
micrinite 23, **26**
micro log 129
microbiological mining 135, 155
microspores 23
middlings: locked 161, **11.2**
migration: primary 34,

34; secondary 34, **34**
milligal 63
mills: autogenous 162; ball 162, **162**, **163**; hammer 161; rod 162
mine detectors 90
mineral deposit 2; alluvial 10, **10**; classification 3; epigenetic 3, **3**; hydrothermal, 6, 8; magmatic 5; pegmatitic 6; placer 10, **10**; primary 3; residual 11, **12**; secondary 3; sedimentary 10, **10**; stratabound 4, **3**; stratiform 4, **3**, **15**; syngenetic 3, **3**
mineral dressing - see mineral processing
mineral processing 159
mineral titles 56, **57**
mineral zoning 7, **7**
Minimata Bay, Japan 203
mining: cut and fill 149, 150, **150**; microbiological 135, 155; open-cut 142-146, **142-145**; placer 135, 140, 141, **141**; pit 142, **142**; shrinkage stoping 150; sluicing 141; solution 135; strip 142-144, **143**, **143**; sub-level stoping 149, **149**; underground 146-150, **146-150**
Mississippi Valley 106
monazite 11
Mount Isa 107
Mount St. Helens, USA 200
Mount Tarawera, NZ **200**
Mountain Pass: rare earth deposits 6
mud: drilling 118; flows 199
multi-bench lateral advance mining 143, **143**
Myponga arch dam, SA **192**
nanoteslas 70
naphtha petroleum 28, **29**, **30**, 31
natural gas 28
Neptunists 4
neutron logs 130
Newton: Law of Universal Gravitation 63; Second Law of Motion 63
nickel: lateritic 11, 12; sulfide 6
nodular bauxite 12, **12**
nonane 30
non-metallic polarisation 86, 87
Norfolk flint mines 135
normal logs 129
North Sea oil 212
nuclear mining 154
nuclear waste disposal 204-206
nuée ardente 199

O/C ratio 32, **33**
offshore drilling rigs - see rigs
oil 32, **33**
oil shales 31, 36
OPEC - see Organisation of Petroleum Exporting Countries
opencast mining - see open-cut
open-cut mining 142-146, **142-145**
open pit mining - see open-cut
Ore: defined 1; dilution 139: indicated 2; inferred 2; measured 2; passes 149, **149**, **150**; proved 2; probable 2; possible 2; reserve calculations 139, 140
Organisation of Petroleum Exporting Countries 36, 210-212
orientation surveys 101
overburden 32, **143**, **143**; ratio 137, 144;

surveys 101
overshot 120
oxidation 108
oxygen isotopes 106

palaeomagnetism 71
palladium 6
Pallisades Sill 6
paramagnetic 70
paraffin petroleum 28, **29**, **30**, 31
paraffin wax 28
pathfinder elements 97
pavements 196, 197, **196**
peat 21 22, **24**, 31
pegmatitic deposits 6; complex 6; simple 6
pelletisation 164
penetration rate 133, 184
per cent frequency effect 83
perched water table 43, **44**
percussion drill 112
permanent saturation **43**, 43
permeability 32, 34, 46, 132, 152, 182; coefficient of 46
perviosity 46, 47
petrol 28, 31, 34
petroleum: aromatic 28, **29**; degradation **30**, 32; engineering 38; extraction 37; gas 31, **33**, 34; maturation 30; migration 33, 34, **34**; naphtha 28, **29**, **30**, 31; oil 32, **33**; paraffin 28, **29**, **30**; quantity **35**; reserves 140, **147**; tectonic setting **15**; traps 37
phosphorite deposits 16
phytane 30
phytol 30
phytoplankton 30
piers 185, **185**
piezometric surface 44, **45**
piles: end-bearing 186, **186**; friction 186, **186**
pillars 146, **149**, **150**
pisolitic bauxite 12, **12**
pit mining 142, **142**
Pitch Lake, Trinidad 35
pivotal pit 143
placer: deposits 10, **10**; mining 140, 141, **141**
plankton 31
plastic: deformation 182; limit 181, **181**, 197
plasticity index 181
plate tectonics **15**,16, 41
platinum 6, 11
Plutonists 4
polarisation: metallic 86, 87; non-metallic 86, 87
pool: petroleum **34**, 36
pore pressure 183, 187
porous pot 82, 92
porosity 33, 34, 42, **42**, 129, 130, 132, 182
porphyry copper deposits: central zone 9, **9** mining **138**, 145; ore zone 9, **9**; origin 8, **15**; phyllic zone 8, **9**; potassic zone 8, **9**; propylitic zone 9, **9**; pyrite zone 9, **9**
potential gradient, 81
potassium feldspar 8
precision 108, **109**
precipitation **40**
preparation 159-164, **161-163**
pressure head **45**
prices of metals, 211-213, **211**, **212**
primary crusher 160
primary migration 34, **34**
primary mineralisation 3
primary recovery 152
primary seismic waves 75
pristane 30
production platforms 151, **151**
productivity 50
prospecting 58

province: geochemical 99, 103; petroleum **34**, 36; metallogenetic 8
pulp 171, **171**
pulsation 167
pumping: water 50
pyrite 9, **9**, 10, 13, 19
pyrobitumen 35
pyrrhotite 70

quarrying 144
quartz 8, 25

radiometric: logs 130; surveys 92, 93
radiation sorting 164, 165, **164**
radon 105
raft footing 186
raise **148-150**
Raleigh seismic waves 75
rank **24**, 25, 32
rare earth deposits 6
ratio: enrichment 177; of concentration 177; overburden 137, 144
reconnaissance exploration 56
recovery 177: primary 152; secondary 152, 153, **152**; tertiary 153
regional gravity gradient 65, **66**
rehabilitation 156
reinforced earth wall 189, 190, **190**
remanent magnetism 71
reserves 1, 138-140; petroleum 140, 209
reservoir: petroleum 34, **34**
reservoir rock 34, **34**
residual: deposits 11, **12**, **15**; gravity 65, 66, **66**; magnetism 70
resinite 23
resistance 80; contact 81
resistivity 80, 81; apparent 81
resistivity survey 48, 80, 83-85, **85**, 183
resources 1, 208
retaining wall 188, **188**, 189, **189**
reverse circulation drilling 122
reverse flow drilling 122
Rhineland coal mining 144
riffles 141, 170, **171**
rigid pavement **196**
rigs, offshore drilling: jack-up 118; semisubmersible, 118; ships 118; submersible 118
Rio Tinto mines 135
Riocin mine **145**
rip-rap 195, **196**
riser 119
risk: exploration 52, **53**
rock fill dam - see earth dam
rock surveys 105
rod mills 162
Roman Empire 135, 203
room and pillar mining 146, **147**
rotary drill 112, **113**, 117-119
rotary table **113**, 118
rotational pit 143
roughers 174, **175**
rutile 11, 166

St. Pierre, Martinique 199
sampling density 101-103
sand 11, 141
sand injection 51
sapropelites 20, 26, 31, 33
saturation: intermittent **43**, 43; permanent **43**, 43
scavengers, 174, **175**
Schlumberger electrode configuration 85, **85**
scintillometer 93
sclerotinite 23
screening 160
screens: well 50, **50**

sealed flexible pavement **196**
seam 20
seat earth **20**, 21
Second Law of Motion 63
secondary enrichment 12, **13**
secondary migration 34, **34**
secondary mineralisation 3
secondary recovery 152, 153, **152**
secondary seismic waves 75
secular magnetic variation 72
seepage 193
seismic ghosts 80
seismic hazards 200-202, **201**
seismic refraction surveys **77**, 183
seismic survey interpretation 76-80
seismic waves: first arrival 75; primary 75; Raleigh 75; reflected 75, **75**, 78-80, **78**; refracted 75, **75**, 76-78, **76**, **77**; secondary 75; surface 75
self potential surveys 80, 91, 92, **92**
semianthracite **24**
semisubmersible drilling rigs 118
separation: analytical - see also sorting
sericite 8
settlement 181, 186, 187
shaft **148**
shaking table 170, **171**
shale shaker **113**, 118
shear strength 182-184, 187
shell and auger drill - see cable drill
Sherrit Gorden nickel mine 164
ships: drill 118
shot firing 51
shrinkage limit 181, **181**
shrinkage stoping 150
sieves 160
silt 192
site investigations, 184, **184**
simple pegmatitic deposits 6
Sinai Peninsula turquoise mines 135
slip surface 187, **187**
slope stability 187-190, **187-190**
sluicing 141, **169**
Snell's Law 75
soil: cohesionless 182, 187; cohesive 182; engineering 180; horizons 102, **102**
solution mining 135
sonic logs 130, **131**
sorting 160; electrical 166, 167, **166**, **167**; gravity 167-170, **168-170**; magnetic 165, 166, **165**, **166**; radiation 164, 165, **164**
source rock 33, **34**
SP logs 130
specific yield 43
spillway 195
spontaneous potential: logs 130; surveys 80
sporinite 23, **26**
spreader **143**
spring 43, 44, **44**
stabilisation 197
stability: slope 187-190, **187-190**; well 50
standpipe **113**
stope **148**, 149, **149**
storms: magnetic 72
stratabound mineralisation 4, **3**
stratiform mineralisation 4, **3**, 145
streak print logging 127
stream sediment surveys 103
string: drill 114

strip mining 142-144, **143**, **143**
sub-artesian: water 44; well **45**
subduction zone 41
sub-base 196, **196**, 197
sub-grade 196, **196**, 197
sub-level stoping 149, **149**
submersible drilling rigs 118
subsidence 156 see also settlement
substitution 210
suction 167
sulfur 19, 153, **153**
sulfur isotopes 106
sump **148**
surface course 196, **196**, 197
surging 50
surveys: electrical 80-92; electromagnetic 80, 89-91, **89**: equipotential 80, 82, **82**; geochemical 58, 59, 101-106; geophysical 58, 59; gravity 63; induced polarisation 80, 86-89, **86**; magnetic, 69; radiometric 92, 93; resistivity 48, 80, 83-85, 183; seismic refraction 183; spontaneous potential 80, 91, 92, **92**; telluric current 80
susceptibility: magnetic 70
swivel **113**
syngenetic mineralisation 3, **3**
synrock 206

tailings 150, 176, **175-176**
tellinite 23
telluric current surveys 80
tensile strength 180, **180**
tertiary recovery 153
threshold **98**, 99, 100, **100**
tidal waves 201
tin 7, 11, 140, 141
titles: mineral 56, **57**
toe **187**, 189, **189**
toluene 28
total head 44
trace elements 202-204
tramp iron 165, **165**
transmissibility 47
transpiration **40**
travelling block **113**
tri-cone bit 118
trommels 160
tsunamis 201
turquoise mines 135

Uinta Basin 31
unconfined aquifer **45**
underclay **20**
underground mining 146-150, **147-150**
Universal Gravity Constant 63
unsealed flexible pavements **196**
upcast shaft **148**
uranium: mining 153

vadose zone 43, **43**
vapour surveys 105
vegetation surveys 104
vertical integration 54
vitrain 24 **26**
vitrification 206
vitrinite 23, **26**, **33**
vitrinite reflectance **33**, 33, 128
volcanic hazards 199, 200, **200**
volitilisation 108

walking dragline 142, **143**
wall rock alteration 8
waste disposal 204-206, **205**
water: connate 41; juvenile 41; formation 41; magmatic 41; meteoric

13, 41; pollution 49, **49**; pumping 50; quality 48; sub-artesian 44
water injection 152
water surveys 105
water table 13, **13**, 43, **43**, 44, **45**, 48, 184, 191, **191**; perched 43, **44**
weathering 14
wedging 151
weep holes 188, **188**
weighted average grade **139**
Weipa bauxite deposits 12, **12**
well: artesian **45**; screens 50, **50**; development 49, **49**, 50; stability 50
well point drainage 191, **191**
Wenner electrode configuration 83-85, **84**
Wenner expanding electrode configuration 83, 84, **84**
Werner, Abraham 4
wettability 171, 172, **172**
White Pine 107
winze **148**
wire-line drilling 120
Witwatersrand 10
worm diagram 106, 107, **107**

xanthates, 172

Yallourn brown coal: briquetting 164; mining 144
yield 48

zinc 7
zircon 11
zone of aeration 43, **43**
zone of intermittent saturation **43**, 43
zone of permanent saturation **43**, 43
zone: vadose **43**, 43
zones: mineral 7, **7**, 8, 9, **9**; alteration 8, **9**
zooplankton 30

For Product Safety Concerns and Information please contact our EU
representative GPSR@taylorandfrancis.com
Taylor & Francis Verlag GmbH, Kaufingerstraße 24, 80331 München, Germany

www.ingramcontent.com/pod-product-compliance
Ingram Content Group UK Ltd.
Pitfield, Milton Keynes, MK11 3LW, UK
UKHW021833260926
475702UK00049B/507
* 9 7 8 0 3 6 7 2 0 7 1 9 9 *